Fuzzy on the Dark Side

Approximate Thinking, and How the Mists of Creativity and Progress Can Become a Prison of Illusion

Fuzzy on the Dark Side

Approximate Thinking, and How the Mists of Creativity and Progress Can Become a Prison of Illusion

Ahmad Hijazi

IFF
BOOKS

Winchester, UK
Washington, USA

JOHN HUNT PUBLISHING

First published by iff Books, 2023
iff Books is an imprint of John Hunt Publishing Ltd., No. 3 East Street, Alresford,
Hampshire SO24 9EE, UK
office@jhpbooks.com
www.johnhuntpublishing.com
www.iff-books.com

For distributor details and how to order please visit the 'Ordering' section on our website.

Text copyright: Ahmad Hijazi 2021

ISBN: 978 1 80341 386 0
978 1 80341 387 7 (ebook)
Library of Congress Control Number: 2022916766

A CIP catalogue record for this book is available from the British Library.

Design: Lapiz Digital Services

UK: Printed and bound by CPI Group (UK) Ltd, Croydon, CR0 4YY
Printed in North America by CPI GPS partners

We operate a distinctive and ethical publishing philosophy in
all areas of our business, from our global network of authors to
production and worldwide distribution.

Contents

Prologue: Approximations, Big Ideas, and Small Perspectives

I like Batman, I must be a genius!

Do you think that "liking" Batman can make you think you understand a tech gadget better?

Very unreasonable, but it – most likely – can!

In researching how some cultural components (resources) can affect the destinies of innovations, I found myself sidetracked toward exploring an interesting phenomenon.

Consumers with different backgrounds were shown ads for two innovative tech products, both non-existent, but one more non-existent than the other. This basically means that one of the products was more complex, novel, and difficult to understand, which was verified by polling a different audience, in what we call a "pretest." There were different ad versions that possible customers saw, and some tried to invoke people's cultural affiliations by using certain symbols and words, while maintaining the same products and a similar message.

We then asked participants to estimate their own understanding of the products, their liking of the products, and the likelihood they'd buy these products.

After controlling for many variables, we found that viewers who saw an ad that had cultural elements they liked (like the Statue of Liberty, or Batman, or Dr Dre) actually thought that the more radical product was easier to understand than peers who saw the same ad but didn't like those elements, or peers who saw a similar ad with no cultural elements at all.

People were approximating their "knowledge" and understanding, even of technical things, as something else: cultural preference.

"I like Batman, so yes... This gadget is not hard to understand."

1

Feelings – affect – (toward unrelated things, like superheroes, celebrities, or the Statue of Liberty) were being mixed with other feelings (uncertainty, for one), and were then influencing understanding (at least perceived understanding) and behavioral intentions.

To assess how much they understood (and liked, and were willing to buy) something, participants were using their love and belonging to a certain cultural group as a guide, even if that group had little contextual relevance. Their own evaluation of how close they were to this cultural group also had an effect on what they thought they knew (understandably).

People do that with brands all the time, and it seems they do the same with cultural belonging (in a sense, isn't strong love for a brand like cultural belonging?). This was one of the goals of the experiment, but similar effects can be observed in different fields too.

Because understanding a technically difficult product isn't a simple task, people were using resources (connected to their self-image and identity) to make sense of the situation and establish a general sense of familiarity. Their own understanding and liking of a novel object was being approximated based on their understanding and liking of something else entirely.

Why? It is much easier that way.

As non-experts, we don't really need detailed understanding, so the approximate understanding, which can be buried under general feelings of comfort and familiarity, will have to suffice...

When we meet new things we meet uncertainty. Uncertainty can be uncomfortable, and people are willing to sacrifice intellectual integrity to get rid of discomfort, or to get things done. The incompleteness of our knowledge is often addressed with different extrapolations and assumptions, sacrificing precision for ease, and reflecting the self onto the world.

This is not always bad, but it can – easily – become tricky.

Our discussions throughout this book will try to expand this perspective. We will explore our use of approximation as a supreme cognitive tool, how it fits within our mental and social/cultural lives, and how it can turn into something dark or problematic. The journey will take us through different fields and (hopefully) unexpected places...It might even have the side effect of occasional clearer thinking and a richer understanding of the intellectual universe.

To make things smoother and more colorful, the book has many quotes on the topics being discussed from diverse thinkers and "influencers," images that visually illustrate certain aspects of the topics, dozens of stories and anecdotes that clarify and illuminate the discussion, and an allegorical short story broken into episodes, with each episode reflecting and gesturing toward relevant themes and incidents that the reader is encouraged to explore and uncover.

Let's look at the two sides...

Meet a fuzzy blade with two (sharp) edges

The bright side: A "good" approximation

The ability to approximate is an amazing mental capacity.

A great deal of our creativity, ability to absorb new knowledge, and problem-solving skills comes from our ability – and willingness – to approximate.

Our ability to navigate complex situations and new spaces sometimes rests on our ability to approximate reality into simple (reduced) systems, with simple inputs/outputs. We neglect some information, and we force the world into our existing (still too approximate) models.

We use approximate understanding to cut through the difficulty and uncertainty.

"Fuzzy logic" is a great example here.

Fuzzy Logic was formally introduced by Zadeh (1965) as a term, but the concept was discussed under the name "infinite-value logic" before, since as opposed to traditional logic, things aren't "True" or "False", but can be somewhere in between. Fuzzy logic seeks to mimic human decision-making under uncertainty and is applied in fields like control and AI.

Think about an artificial intelligence module trying to park a car. When you (a human?) park a car, you don't need exact awareness of all the relevant variables and measurements in all directions and at all times. The information overload will be overwhelming, not to mention wasteful. Many variables can be approximated and neglected as long as they are within certain boundaries. Approximation here helps us (humans and parking artificial intelligence systems) get the task done efficiently with a reduced set of completed calculations.

Think of an artist creating a revolutionary work of art. They probably work with many unknowns and summon symbols that bridge different (mutilated) disciplines to make their statements. They work with incompleteness most of the time.

> *"I'm no expert mind you. I steal whatever I see without knowing what it is."*[1]
>
> Pablo Sorrentino

Scientists researching new fields try to construct simplified models that approximate the real scenario, by reducing reality and "controlling for" many different variables.

A great business manager or entrepreneur will have a certain (illusory or unrealistic) vision of a future where their creation becomes a strong force in the market, and they construct their forecasts as approximations based on estimates of a few variables that have been defined (somehow) as the most relevant.

In all these scenarios, using approximate understanding helps us shape and create a new reality against the odds of the game, rigged by so many unknowns and the overwhelming incompleteness of knowledge.

Engineers (and physicists and mathematicians) are happy (it is said) to assume a horse is a sphere (to a biologist's horror, I think) if needed. This approximation is acceptable because it helps solve a problem! (I will not even begin to talk about the sins of economists...for now.)

In all these examples, the approximation is a tool for progress. It helps solve problems and create new realities. That, however, isn't always the case.

The dark side

There is a caveat: which problem is being solved by the approximation? Are you aware of the approximation as you commit it? What are you approximating "away"?

The answers to these questions – among others – are quite important, as I argue in the coming chapters.

Because of this tool's potency, we sometimes are inclined to overuse it. We start applying approximations in fields in which they are not needed, or worse yet, we use approximations unknowingly and iteratively, and drift – like a sailor listening to a siren's song – losing track of the actual problem.

This is one of the themes of this book. After looking at why it is inevitable that we use approximation as a mental tool, by considering both reality (the world), and the observer (us), I present applications (and circumstances) in which approximation causes more harm than good, showing how it can be a trap leading to a waste, instead of conservation, of resources (mental and otherwise).

I've personally seen this many times.

Periodically, as I observe arguments and discussions happening in front of (with) me, I find that they are – almost – completely caused by an unconscious approximation that one of the parties is doing (willingly or not)...I encourage you, dear reader, to do the same (observing, not unconscious approximating), and I think that after reading this book it will become more evident.

In the coming pages, and after trying (sufficiently) to understand the general theme, I will try to explore the phenomenon in different areas, ranging from personal decisions and emotional/rational behavior, to management, politics, thought, science, and language.

Perspective

The book is presented like a chain of short thoughts and discussions, and I intended to make it as simple as possible, at the expense of (some) academic rigor. I aimed for simplicity, even though text analysis AIs don't think I've done a good job, and insist that many sections are "too difficult" (could it be a faulty approximation?).

The discussions in the coming pages explore a set of recurring themes. They are mainly organized around approximation because approximation is the most concrete (ironic, I know) and practical concept in connecting these different fields, and applies to understanding most naturally and directly.

That being said, this is a book about better thinking. The themes of incompleteness, creativity, ignorance and laziness, identities (cultural and otherwise), self-awareness, systems, and complexity are all part of the same picture that I try to construct...and contemplating their effects was one of my motivations for writing this book.

Figure 1: An approximation of the set of secondary themes

The book is also – quite importantly – about progress and the impediments in its path. The progress of the individual and the group toward a richer and more satisfying state of affairs: Progress – as I argue in the coming chapters – not as a means or transient state, but as a destinationless goal and governing theme of existence. It is not the destination that should concern us the most, it is the conscious and intentional effort that we continuously dedicate toward being on the path.

For the sake of simplicity, I've omitted academic-style hypothesizing, and tried to reduce jargon as much as possible. I like to think that it is also interesting to explore how this central concept has implications in many different (and disparate) disciplines, and how it has direct effects on our everyday personal and professional lives.

Mixed thoughts and feelings (and a disclaimer!)

There is something "freakonomics[2]-y" about what I'm trying to present here, as the coming pages carry many stories, musings, frameworks, and ideas from different disciplines. I promise, however, that there is a clear and meaningful on-point thread that connects all this[3], and it goes beyond what we will directly discuss (a synthesis, of sorts).

The journey should be entertaining and informing, and present an opportunity to look at different areas and fields tied to psychology, organization, systems, management, and communication. Looking at what connects many disparate fields has always been in my hidden agenda. But beyond that, I hope that the knowledge acquired from the coming pages can contribute to overcoming the many prisons and traps that are placed everywhere by masked approximations (i.e. ignorant incompleteness, we'll talk about that...when you're ready!).

Reflecting on approximation should pave the way for clearer, and more effective, thinking.

I don't claim that the coming pages present a theoretical breakthrough, nor do I faithfully preach a dominant model.

My aim is too humble for that. I try – instead – to introduce a practical – and potent – perspective (a super idea), and to establish some kind of order as I organize and link different concepts through its lens.

Perspectives and Super Ideas are important concepts here. I hope you will see – as we go on – how "approximation" emerges as a "super idea" capable of organizing different themes and explaining many (seemingly) disparate phenomena...A distillation into a higher truth.

I'd also be content and thankful if the discussions in the coming pages encourage an examination of completeness, uncertainty, identity, creativity, and our interactions with knowledge. In doing so, they might give a fleeting glimpse of very elegant interconnectedness.

But first, let's meet Zif...

* Zif's Destiny

(Or: A (seemingly) non-terminable task)

Grumis saw the man who was completely focused on what seemed to be an endless task.

He was going up and down the scaffolds of the incomplete structure, carrying random rocks to the higher levels. There were others doing similar tasks. The job was futile, however, because the workers constantly dropped rocks as they worked. It didn't seem that they were building something consistent, so parts of the structure kept crumbling occasionally.

The man didn't seem to have time to stop and assess what was happening, as he toiled loyally and religiously. Grumis was here to learn, and his process detectors indicated that this was a high-priority learning opportunity.

The green robotic bird approached the man.

"Good day! I am Grumis-33, and I'm in the third learning phase. Do you mind if I observe and ask a few questions?"

The man looked at the small creature, and answered in a friendly manner without ceasing his activity: "Sure! I don't mind some company. Other than the workers who help me, I haven't talked to anyone in ages. Call me Zif. What do you want to learn?"

"What are you building?"

"A tower. The mountain tower."

"When did you start?" Grumis asked as he registered the new information and searched the databases he could access.

"I'm not sure anymore. I've been working on this for as long as I can remember," Zif explained without stopping his rock-moving task. "For some reason, it doesn't seem to be progressing as it should be."

"Why?"

"I'm not really sure. I don't have time to figure that out exactly. There are too many people working with me, many tasks to complete,

and just too many ideas. Sometimes you can't understand a thing these workers are saying – and I think that many don't understand each other too...It seems to me this gets worse as our work progresses. Maybe if you fly high enough you can see better and tell me if you can spot any problem I can fix?"

"I did on my way here. I couldn't tell exactly what was happening, or if there was something wrong."

"Yes. Me too. Just need to continue building this tower. The materials we have aren't exactly made for the task either. All the rocks are so irregular, and this seems to be the cause of many mistakes and delays. The workers try to fit them as much as they can, but it doesn't always work."

He docked a couple of falling objects, and continued his infinite task of hauling rocks up the incomplete structure.

The workers – and Zif – were doing a relatively good job of coordinating their work, given that they didn't seem to talk the same language. They were surprisingly creative in getting the irregular stones to fit into the structure...although it was clear to Grumis that they wouldn't easily get the task completed.

"Do you mind if I stay here? I'd like to learn more about this tower of yours."

"Not at all."

Notes

1. The quote is from Sorrentino's beautiful film *The Hand of God through the Eyes of Sorrentino* (2021).
2. "Freakonomics" is a best-selling book by Levitt & Dubner (2005) with many surprising observations and analyses.
3. In stressing the difference between what I'm trying to do here and the great work in "Freakonomics" I felt that I was – eerily – mirroring Cartman's (from *South Park*) rant, as he explains why he's not like Family Guy (from *Family Guy*).

About Approximation: Attempted Clarification

The quest for certainty blocks the search for meaning.
Uncertainty is the very condition to impel man to unfold his powers.
Erich Fromm

Accuracy is overrated (realistically speaking)

Approximation is everywhere around us, and humans are – by nature – not accurate creatures (intellectually at least).

We use approximations whenever we can, and then some more.

Consider the following three areas:

- Approximations in the conceptual field
  As we try to understand new ideas, or even as we ponder existing concepts, the richness of the intellectual scope generates many unseen approximations and associations that can either open new doors or blur understanding.
- Approximations in the communication field
  These are quite obvious, and they are approximations that relate to how we communicate our inner worlds to others (or how we understand theirs). The whole process of communication is riddled with approximations from all sides (the communicator, the listener, and the medium). The undesired components are called "noise."
- Approximations in the actions field
  We frequently do something with the intention of achieving a certain goal, but our actions rarely align with our goals. The real/social world is just too complex for that to work. This failure can happen at both individual and group levels, with institutions, as well as people,

failing to achieve goals for a variety of reasons stemming from unseen approximations.

Take interpersonal communication.

You've surely noticed – dear reader – that we (mostly) don't say exactly what we mean to say. Approximation lives in all stages of the communication process, and this has been the subject of significant research. We say something that we think reflects our ideas (it doesn't accurately, we – after all – have emotions and aren't infinitely intelligent), and we assume that our message will be received by someone as we know it (it frequently gets distorted), and we assume that the other person(s) will understand what we meant (they frequently don't – they have emotions and their own ideas and prejudices too).

In the simple act of talking to someone there are approximations in expressing the idea, in transmitting the idea, and in receiving/understanding the idea.

But it's not just a communication failure. Our thoughts themselves lack accuracy.

We understand new concepts based on our existing understanding of other (old) concepts. We effectively "borrow" characteristics from one context to another.

We make decisions based on what we think are the most relevant criteria (this is what a rational person would do), but we are in fact simply reacting to our approximation of a big set of variables.

Bounded Rationality is a concept proposed by Simon (1955), saying that people don't always consider a full analysis prior to making a decision, but due to limitations on cognitive capacity, skills, time, and other resources, make an optimal (satisfactory) decision. More on this later.

It doesn't even stop there!

Political leaders are (supposed to be) approximations of the collective will of the population. Languages allow for only approximate expressions of some concepts (they were designed to be practical). Scientists think they are going after the truth, but they are – in fact – going after a distorted (yes – approximated) version of the truth that has been veiled by the specificities of their field and perspective.

Approximation is a universal principle that exists in the physical world, not just in our conceptual one. The electron, and every other basic-level component of matter (if such "things" even exist), are only "approximately" there. They're much less stable and fixed than you think. Think about the edges of physical objects – they don't really exist either. These (seemingly clear) edges are simply the point at which the repulsion forces between approximately-there electrons become strong enough...but let's go back to the intellectual realm.

As we track approximation as "satisfactory replacement" we'll go through some of the different fields I've just mentioned, but first we need to look more closely.

Elements: the accurate, the useful, the error!

I don't want to dedicate too much space (time) to general theorization, but let's say that approximation – most simply put – is substituting something (the approximated – or object 'A') for something else (object 'B'), because using B is somehow more convenient (details later).

So in some context of thought or communication or action, object A is replaced by object B.

In simple terms, approximation is using (thinking/communicating/doing/...) something (B) instead of something else (A)...

Keep in mind that B is not exactly A. They are *almost* the same, but not quite.

We choose B – specifically, rather than C or D – for many reasons, and they could be cultural, personal, cognitive, emotional, etc...There are differences between B and A: E=A-B. If you're luckily using an approximation (or possess unlikely levels of mental presence), these differences will be negligible in your specific field of application (so, E ~ 0).

Sadly, they frequently aren't (because let's face it, you don't possess unlikely levels of mental presence).

So approximation:

$A \rightarrow B$ (We use B instead of A)

$A - B = E$ (E is the difference between the two)

$E \sim 0$ (E is negligible, so the approximation is fine)

Problems start when E is non-negligible.

This concept of error is actually important in engineering (telecommunications). An error between a received (Rx) and a transmitted (Tx) signal is constantly monitored and is usually required to remain below a certain threshold to ensure the reliability of the communication channel. The more predictable (systematic) the error is, the easier it is to eliminate and maintain better transmission.

What I want to do here is apply this concept to different personal/social/interpersonal situations, and observe how mischievous this "E" can be when left to its own devices, unchecked and uncontrolled for.

In doing this, I try to apply concepts from different fields, ranging from engineering and math to systems thinking and management, within social and psychological contexts.

I want to champion a predictable and non-surprising (less entropic) E!

Here are some examples of approximations and their errors, to make the above discussion more relatable (as if that was possible):

Table 1: Examples on Approximated Issues: Decisions and Thoughts

A (an object. The Accurate)	B (what it is replaced by. The Useful)	E (The Error)
The will of the people with regards to government and legislation	The will of elected members of Parliament	The difference between what the politicians do and the people want
My ideas about approximation, thinking, management, and creativity	This book	The result of my inability to express the ideas clearly enough
Zaha deciding on her university specialization, based on what is best for her	Zaha deciding on her specialization based on her projection of what she thinks her deceased parents would have wanted her to do	The difference between what Zaha really wants (which fits her personality), and what she thinks her parents would want

Thinking in terms of "The Accurate, The Useful, and The Error" in different contexts can lead to clearer understanding.

Non-random errors are friendly

E in this context can be any difference or distortion between the real item (thought, message, action, intention) – A – and the approximating one – B.

Things get more interesting when the selection of B is not entirely random, and so E is not entirely random. The error is usually a residual cultural or emotional (psychological) or personal or historic set of components, that relate to the particular situation in which an approximation is occurring, or to the person doing the approximation.

Let's assume I understand a certain issue incompletely, and – in my mind – some of its components are approximated. I may be aware of this approximation, and I may not. The error resulting from the approximation depends on my existing knowledge and preferences, because this is where my set of analogies comes from, and it might depend on my cultural background, because many of our values and meanings are connected to our cultural affiliations.

So why is this good?

Well, it is not that good, but it is – at least – good for the purposes of the discussion in this book.

It goes back to the problem of removing the error. If the error is predictable, it can be more easily seen and neutralized (versus the case when it is completely random). If you observe your own approximations, with time, you will discover that certain patterns of errors emerge (in your thinking or communication), and they become easier to identify, precisely because they share common characteristics.

The bright edge of the blade...the dark edge of the blade

Approximation has replicated itself in human society because it is quite a successful and useful cognitive tool (to use evolutionary language).

It isn't very hard to imagine that if there was a gene for approximation it would have a killer evolutionary advantage. All the humans without it (the less-known, now-extinct, *Homo Pedanticus*) would tragically perish as they try to understand exactly what that yellowish object with sharp face-tools approaching them at a high speed is.

Approximation presents many benefits and is in fact inevitable when we consider the complexity of the environment and the nature of our interaction with it.

Approximation allows us to reach out into new areas of thinking, and is closely related to our scientific progress, because all scientific models are approximations of reality consistent with our existing knowledge up to a certain point. Our whole scientific endeavor as a species can be summarized in a continuous search for better approximations...and this is true even with exact sciences. Newton's description of the gravitational force ($F = G \times m1 \times m2 / r^2$) is an approximation of reality, Einstein's improvements are better approximations, and quantum mechanics are good enough approximations of reality considering what we know now...They are all good enough approximations within certain limits.

Approximations are very practical.

"Exactness" and "Precision" – keep in mind – are not absolute values. They are needed only to a level sufficient to solve a particular problem. Approximation allows us to get to results without needing to accurately solve some secondary problems on our path to a specific goal (these secondary problems are "approximated away"). This is an excellent example of "constructive ambiguity."

This is fuzzy on the bright side!

But we're prone to overuse tools that work.

In addition to being inaccurate, it seems we have a tendency to overreach (another great tragic human flaw, or blessing).

We get used to our great mental tool/habit, and suddenly everything becomes a nail.

It takes too much effort to dissect many issues so we get accustomed to approximating them away. In doing that, we replace genuine effort that can lead to growth and development (on individual and group levels) with murky shortcuts. This is when clarity is possible and attainable, but somehow "hard to reach."

We start relying on mysterious emotional inclinations without contexts, instead of proper understanding, for solving problems or making decisions.

We surrender to inertial forces that make us stick with some dated social or political or managerial institutions that can be replaced, or even let go of a certain quest for truth (scientific or other) without trying to estimate how much more we can advance, because where we are is "good enough."

Sticky and persistent illusions can result from approximation errors as concepts get entangled in a web of – sometimes non-existent – relationships. Inefficiencies and complacency can thwart well-meaning efforts, and inertia can slow the advent of needed change when "problem soups" dominate. They can even lead to the escalation of perfectly avoidable conflict.

Approximation can become dangerous for individuals and groups, and this is what I mean by the "dark edge" of the fuzzy blade (I am sincerely sorry for letting you – the reader – struggle with the idea of trying to imagine approximation or fuzziness as a blade, and with two edges, no less).

Even though it can basically enable them, approximation can stand in the way of true understanding, true thinking, or better action/management.

The coming pages will explore these concepts, and I encourage the reader to think with an open mind about some of these problems, and what it would really take to resolve them on a personal level.

I will discuss some prescriptive ideas toward the end of the book, but I believe that simply using the "approximation perspective" to become aware of what is happening on psychological, social, cognitive, and cultural levels is a great step toward resolving many of the risks!

Ambiguity is a prerequisite for creativity. Deal with it.

An idea is a point of departure and no more. As soon as you elaborate it, it becomes transformed by thought.

Pablo Picasso

The discussions in this book about approximation in thought, speech, and acts leads to the contemplation of "clear thought," "clear speech," and "clear action," and how feasible they are. It is very important to note – from the beginning – that such things are unlikely. They are ever elusive goals, and this is part of the nature of life and human activity.

A brave embrace of uncertainty is a prerequisite for creativity. Creativity depends on our ability to cope with ambiguity, and our courage to explore areas that aren't clear. We have to use analogies and approximations extensively to keep discovering the world around us, and to create new – conceptual and mental and material – worlds.

Seeking absolute clarity – at the expense of mental courage and exploration – can lead to extreme "incrementalism," which I discuss in one of the coming chapters as a serious problem in itself, where the bigger picture is replaced by (approximated away for) a few steps forward (so an approximation is replaced by an even bigger approximation later on).

That being said, some approximation and ambiguity doesn't lead to creativity and progress, but rather is capable of preventing us from exploring new frontiers and going beyond our existing prisons. This is why I think that the image of the double-edged blade is opportune. Being aware of approximation occurring might allow for a better interaction with different ongoing social (and mental) processes, leading to more clarity, hopefully without being too incremental or excessively analytical.

A (premature) synthesis

The next diagrams show some of the elements surrounding the processes of approximation. These elements will be explored before we move to applications within the fields of science, language, markets, management, and more.

The incompleteness of our knowledge, the complexity in our environments, and the limitations of our mental resources mean that our understanding, communication, and actions (individual or collective) are usually approximations of some ideal. This is especially true because our identities are tightly interwoven into what we think, say, and do!

This "approximate" quality is itself value-neutral. It has the potential to be constructive and progressive, or regressive and fossilized. Its outcomes are subject to many factors, and hence, "luck."

But that's not all – it is also subject to our own knowledge and experiences: these determine the set of all the possible "Bs" that we use (approximating concepts, actions, or statements). The outcome of actions also hinges on our awareness of the ongoing approximations – this awareness opens the door for easier and more frequent improvements of the models we invoke.

We ordinarily have little conscious control over this, and this doesn't have a trivial solution. Because you can't just pay attention…You have to become the person who pays attention.

As a rule of thumb, more openness and awareness, a humble attitude, and a broader experience (or knowledge) usually translate into more fortunate encounters with approximation.

The Brighter Edge of the Blade

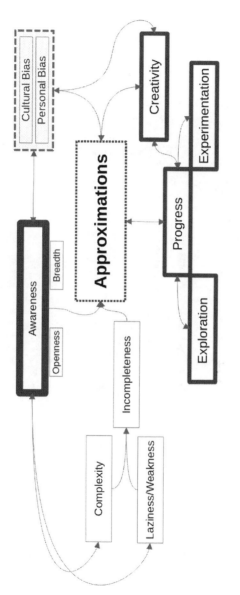

The Darker Edge of the Blade

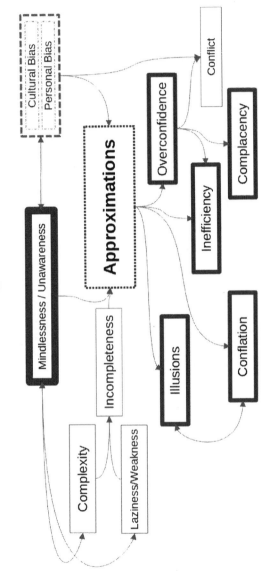

Approximation is a good approximation

So this book is not about presenting a ground-breaking invention (if this is disappointing, sorry). The effort presented here is a collection of ideas and observations that aim to establish a "perspective." This perspective – if used – can help clarify thought and explain why our most sincere efforts sometimes fail. It is also about living with complexity.

Being somewhat of a wanderer, I've studied (and taught, and worked with) concepts in different fields including markets, innovation, decision-making and biases, engineering and IT, cultural industries and resources, social networks, consumer behavior, and leadership. What has been particularly interesting to me as I went through these disciplines was the emergence of common themes that can be applied in different contexts (the boundaries of scientific disciplines and fields are – after all – somehow arbitrary: where is the boundary between biology and chemistry, chemistry and physics, physics and cognition, psychology and sociology, etc...?).

Approximation was always among these recurring themes. Many disparate and seemingly unrelated issues we'll explore can be understood more easily in terms of our tendency to approximate. I sometimes examine a polarizing idea and find out that a conflict can – simply and surprisingly – be resolved by trying to understand what is being approximated (replaced by an inaccurate representation), and the error that results from that.

Two frameworks from management are relevant here.

The "5-Why's"
The first is the 5-Why's framework, which is part of Toyota's lean management set of tools. It is used to find the root causes

of a certain problem and the flow of actions and thoughts that caused a certain issue. The technique (shockingly?) consists of asking "Why" five consecutive times starting with "why did the problem happen?"

Asking the 5-Why's to understand conflicts and failures in interactions can lead to the identification of assumptions (with varying degrees) made by one party or the other. These assumptions are usually the root-cause of a clash or block, and they sit at the heart of unseen approximations.

The 3 levels of culture

The second relevant framework is the work on organizational culture by Edgar Schein.[1] Schein says that organizational culture can be thought of as having "levels," starting with a visible and tangible level, that of objects, rituals, and behaviors – the level of artifacts. People see and interact with artifacts daily, and artifacts shape their experience with culture. Artifacts are a culture's material manifestation. The next level is that of values – these are expressed and conscious statements of what is valid and good and what isn't. The fascinating and mysterious third level is the level of implicit assumptions. What do we "just assume" is right? Assumptions are frequently unnoticed and they give the rise to the other levels. (So you can like the flag (object/artifact), because you are patriotic (value), because – maybe – you assume that having order and large grouping is good and natural (assumption)).

Asking the 5-Whys can lead to the identification of an absent, sometimes rogue, approximation. Thinking in terms of what is hidden beyond what is said and done can reveal important assumptions that are the cause or the explanation for the error (E): the difference between what is intended (A) and the approximate replacement (B).

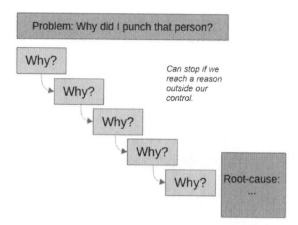

Figure 2: The 5 Whys
Note: there might be more than one line of reasoning.

The super powers of super ideas

No one phenomenon in the social/psychological field can be thought of as dominant, or one theory as having extraordinary explanatory power regardless of circumstances. The scale of complexity is just too vast.

Some ideas, though, are in a class of their own. Their explanatory powers extend across a range of fields and disciplines. They are elegant representations of abstract truths that can gracefully explain things in different contexts. A great example is the idea of evolutionary logic, and how it applies to a wide range of topics from biology to sociology to economics to management (market competition, innovation, etc...), offering insights in all these fields.

In this sense precisely, the approximation perspective is simple and parsimonious. It is simple enough that it can be remembered and applied easily (do you know how many cognitive biases there are? Read on), and parsimonious in the sense that it can be enough to resolve and explain many issues. It is minimal and powerful, and if you're intent on reducing the number of concepts in your mind (believe me, you probably are! [2]), or at least link them better, having these kinds of ideas (simple, powerful, cross-disciplinary) really helps.[3]

In other words, the approximation perspective is a great approximation (I'll talk about recursion...soon) for many of the problems that we encounter on a daily basis (outlined in the coming pages). It is a "super idea" that can explain many things in different fields. It is for this reason precisely that I dedicate some effort to defining its components and trying to establish a somehow formal understanding in the upcoming (annoying – for some) chapter.

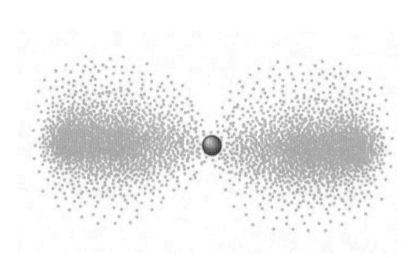

Buzzwords and frameworks are the lesser evil

I remember a lecture once, where in the middle of a heated debate with a professor, I quoted a statement by a popular psychology author, and he dismissed the discussion simply by saying: "(that guy) is a buzzwords' guy." "He just embraces a few buzzwords, and uses those for everything," he continued.

After all these years, I can't say that I've become an enemy of buzzwords, even though I try to be an anti-reductionist (I admit that I do that when convenient, like any self-hating reductionist). Evil as they can be, I don't see buzzwords as an entirely bad thing. We merely have to qualify the approximation (a buzzword is a popular approximation of an underlying theory or set of ideas).

In my professor's mind, buzzwords are a replacement for hard work and strong theoretical foundations. Buzzwords are synonymous with "cheap" publicity. I think – however – that sometimes buzzwords can be helpful. Buzzwords can help us remember things, and make us look in a specific direction (even when our distracted nature wants us to look everywhere).

This relates to a problem we have in business education too. The strong compartmentalization of business education, which can cause short-sightedness in the long run, or seed students' inability to synthesize all the needed information or to look beyond their immediate functional areas, is an ever-present concern. One of the methods that can be used to combat this is to caution students against overusing a certain framework (approximation), or to use different frameworks to analyze a situation from different angles if possible. I always caution students that a framework is – to simplify – just a limited perspective. It is like a filter that shields you from the information (or data) overload. This is dangerous, but it is also powerful. In helping you focus your attention on specific variables it can sometimes reduce the required processing and difficulty of a problem, while guaranteeing a certain degree of comprehensiveness. It is not the best tool, but it can be a good – and convenient – place to start.

I claim that the "Approximately" perspective is – in a rough sense – just like that. It can be a simple and easy way to deconstruct a conflict or assess a problematic situation. Go into more detail later if you want, and invoke deeper analysis, but looking for the hidden approximation can be your preliminary Occam's razor (Occam's razor or the law of parsimony roughly means that the simplest explanation is usually the best).

"When you hear hoof beats, think: 'horses', not zebras."

Dr Perry Cox (*Scrubs*)

* Zinbiela's Ledger

(Or: A good idea...Till you try it)

Grumis tried to evaluate the situation and probe the curious character he just met for more information as he tried to understand. He was not allowed to instruct or advise the characters he met, and that was a fuzzy line that he had to mind.

"It seems that you are facing quite a challenging task," he said.

"It is what needs to be done."

"Still, did you try to do things differently?" It seemed that there were many things that needed to be done instead of this incessant rock moving, if the goal was to build a tower.

"Yes. We tried many different things. It isn't simple, though. We learned the hard way that no matter what we try to change, we will face problems. The problems we kept facing were always something we couldn't have predicted. If I didn't know better, I would say that there was something magical about this task," Zif complained.

"How so?"

"Well, you might think: Why didn't you try to fit the right stone in the right place to build something more stable? I once tried to get someone to survey every possible stone here on the land and design the tower according to what we have. It is a giant task, but seemed easy to my innocent eyes at the time. Zinbiela tried to do that, only to discover later that the stones she surveyed and took record of...didn't exist. It is like they kept changing...At first we thought it might have been her fault, and I double checked. We surveyed, but after a while, the stones really were different. Things you recorded didn't remain as they were. Was it because they were recorded? Was it simply their nature? We couldn't tell. They are so many, and so tricky!"

"Strange!"

"Yes. Creating a plan was like saying that we needed to put a bell on the cat. Sounds great, doesn't work."

"Did you try something else?"

"Sure – surveying the stones wasn't the whole story. We thought of giving the workers specific tasks - you know, telling each one exactly what to do instead of just instructing them to build. But that was a huge problem. We couldn't even begin to understand the scale of tasks needed, let alone get everyone to do their portion. You can't – it turns out – break things down to any level you want. The more tasks you specified, the more tasks you had to specify – it seemed infinite. We quit that attempt, and tried something else. It seemed more promising at the time."

"What was that?"

"Well, a group of workers were told to try and build a detailed model of the completed tower. Their model was brilliant. We started with a 1000:1 scale, which was quite detailed, but soon problems started when the workers disagreed on what needed to be done in certain places where the model didn't give enough information. We naturally updated the scale and made it more detailed, and so we got a 100:1 scale, which took years to construct, and was truly gigantic in size. Our model had a much higher resolution now, but we still faced problems – what was I to do next?"

"Build a 10:1 model?"

"Well you might think that's reasonable, but what if the same issues came up? Would I do a 1:1 model of the building? The 1:1 is the only model that can resolve the possibility of disagreement. But that meant that we needed to survey the stones (which didn't work), and to give workers specific instructions (which we couldn't). Do you see where this is going?"

"Yes. I see now...So what did you do then?"

"What I've done for as long as I remember: continue building the tower of course."

Notes

1. Schein's (2006) work on organizational cultures has great insights on the structure of cultures as systems of assumptions, values, and expressions.
2. But if you can – dear reader – fight the tide of reductionism, do it! I'll cheer for you.
3. If you're interested in more examples (other than evolution), of these "super ideas," consider the cross-disciplinary potential of quantum mechanics (physical, and metaphorical-philosophical levels), artificial intelligence (computation, mathematics, cognition, philosophy of mind...), emergence (biology, cognition), etc...

A Formalization: Approximation as Relaxed Boundary Conditions

The three common paths of reduction

Usually, we approximate one thing (A) as another (B) because we assume that E (E=A-B) is negligible (almost zero), or irrelevant.

In effect, sometimes A and B are neighboring items, and sometimes one of them is more general, and it gets approximated into a subset (part) of itself, because its remaining characteristics are considered to be irrelevant or completely overlooked.

So when we approximate we are (generally) replacing:

- a whole with one or more of its parts,
- a part with a whole (set) that contains it (and other things),
- a neighbor with a neighbor (or parallels).

Let's think about the racist person (more on this nice species later).

He/She typically assumes that one particular individual (real or typical/metaphorical – the "stereotype") belonging to a certain group is a suitable representation of the whole group (sometimes in terms of good qualities, and sometimes bad ones). Someone might tell you: "X-people are really hard-working," or "Y-people are thieves." In effect there are a few approximations (not one) happening here:

- The big group "X-people" is reduced in this person's mind to being hard-working, and being only (or mostly) hard-working. Thinking of them as having one (or too few) trait(s) is convenient [part-for-whole].
- "X-people" have been approximated as hard-working based on some anecdotal evidence (which is an

approximation for proper evidence) like "John is an X-person. He is hard-working, so X-people are hard-working." [part-for-whole]

- This might be taken back to individuals, so "John is an X-person, so he is hard-working, because X-people are hard-working" [whole-for-part] → This is a slippery slope!

The mistakes typically are part-for-whole, where we approximate the whole based on a part ("John is a good representation of X-people," "I know X-people because I know John," or even "X-people are lazy, and not other things" – if in fact they happen to be lazy), whole-for-part, where we approximate the part based on a bigger set ("People with PhDs are smart and successful"), neighbor for neighbor, where we approximate one object (group, person, item...) as another because they are "somehow similar" (e.g. "Our nation is a nation blessed by God": In saying this we mean that our nation stands for our group loyalty and righteousness and history and heritage, and the texts that we have, which tell us about our God, insist on these values).

This replacement is often used in all situations where we don't have enough information or resources in general to consider an issue accurately. As we saw, this is a great skill that can save us a lot of work and effort, and allows us to neglect many irrelevant variables.

Research in communication[1] describes something similar when talking about relations of "equivalence" between statements and objects. Sometimes incorrect relations of equivalence (i.e. approximations) originate because of a certain particularization, or a generalization, or a parallel (aligned) structure of the two.

On loss and distortion: approximations as projections into other spaces

I sometimes visualize an approximation that I think about as something resembling a geometric projection, and I think that this mathematical operation elegantly captures the essence of what I'm trying to discuss here.

A geometric projection is when one object is cast (projected) into a space other than its native one. Sometimes we can project an object while retaining all its dimensions, and sometimes we project the object to a space with less dimensions for convenience (think about drawing a cube on a sheet of paper), reducing our total information load.

Here are some examples of projecting objects:

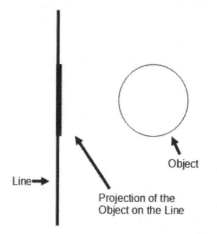

Figure 3: Projecting a circle onto a straight line

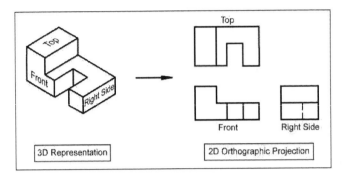

Figure 4: Different projections of a 3-D object onto a 2-D plane

Approximation is just that: a projection. We move one object from its "original" location and fit it into a new space or a new perspective, sometimes with less (or skewed) dimensions (A → B). The projection into the new space sometimes takes away information. Interestingly, projections into a space of the same number of dimensions can cause distortions too, depending on the perspective. A square in one plane might be projected into a segment (line) in another plane, if the two planes are perpendicular (extreme case).

In a sense, we are always slicing reality according to our positions. The slices are what we experience, and they are always a simplification of some richer picture.

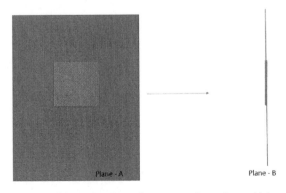

Figure 5: The projection of a square (on plane A) is a line
(on Plane B)

In this image, Plane B is perpendicular (so think of it as going into the page, and coming out of it, at a right angle), and looking from that plane, at the side of Plane A, one would see the square as a line. Here both spaces have the same number of dimensions, but the distortion has been maximized, and the square is completely distorted into a line (loss of a complete dimension). If we were projecting the square into a line, it will always be a line (the target space allows only for one dimension).

Approximation, in shifting the object from one reference to another, can lead to similar distortions either in understanding, or in communication, or in unintended consequences of action and organization.

Note that we are always constructing the world mentally based on sensory inputs we receive. We are always projecting.

As I tell you something, for example, my words can have different connotations and might be embedded in different meaning networks for you. They can create more (or less) thoughts when you hear them compared to my original intention.

The insight here is that a specific projected (approximated) design or situation might correspond to an infinite set of possibilities if the reference frames are not aligned. Our set of selected approximations (the set of all the "Bs" we use, which is probably a variation of our past and idiosyncratic experiences) can be a source of infinite distortions and chaos if it happens to be removed and catastrophically un-aligned with a more universal reality (the set of approximated objects – "As").

Throughout our lives, we constantly shift objects between perspectives, ours and those of others, and usually with complete disregard to how distorted by emotions, loaded meanings, personal experiences, and cultural biases all these projections and shifts are. Sometimes, even with the awareness of the approximation, the information used is not enough to reconstruct the original (just like what happens when you lose a dimension).

The book *Flatland* (by Edwin Abbott, published in 1884) is a really inspiring short novel, and I'd recommend it if you're interested in thinking deeply about these issues of dimensions and how perspectives can totally distort the nature of reality.

A (sad) story with transistors

As a naive first-year engineering student studying transistors, I remember the moment of discovering that I was doing things completely wrong when I (accidentally and half-heartedly) joined a group of friends and colleagues studying for a midterm. As I saw how everyone was solving problems, I realized that I had spent too much time and effort trying to understand things down to the molecular level, reworking some formulas and shortcuts every time. This was causing the loss of precious time when I tried to solve problems. I wasn't using schemas in the right way, and even though my understanding could have been more "fundamental" and "grounded," it was a source of confusion and waste.

My work wasn't "reduced" sufficiently.[2] If this weren't engineering, I might have been okay. But ironically, I was really doing the anti-engineering thing. Efficiency was the game, and I was stuck (in the wasteland between engineering and physics).

I fondly remember a colleague later telling me that he wanted to shift majors, because he didn't see himself several years down the road "worrying about a traveling electron"...I agreed then, and that was all about trying to discover our paths in life. Still – I wonder – who should worry about traveling electrons?

To be perfectly clear, there is so much beauty and intrigue in digging deep and trying to peek a look at the fundamental governing rules of the universe. But that won't let you solve 20 problems in an exam.

The moral of the story is: you need to be aware of which level (or layer) you're working on if you want to avoid getting lost. There are many levels and planes that we constantly move on. It is not that one level is wiser or better or more objectively relevant. It is about understanding your own place, and trying to solve the problem at hand with the right eyes and the right tools (this is good engineering thinking).

Your perspective, and your awareness of it, are elements of your toolbox.

Ken Wilber, in many of his books, introduces very elegant ideas that help stress this, and the need for understanding the level on which you (the subject) or the problem you're working on (the object, it) stand (here I'm taking some liberty in extrapolating these ideas). Different "stages of consciousness," for example, exist, and each one has more complexity and is inclusive of all the previous ones. These different layers of reality are sliced projections. Without conscious attention, we get blinded and think that our layer is the whole thing...

You can work with the line on one plane, and in fact you might need to sometimes. You should still remember that from another person's perspective, this line is a circle!

3

Notes

1. The research by Lewandowska-Tomaszczyk, Barbara (2012) is an example.
2. Reductionism is discussed in greater detail in the chapter on Science & Approximation.
3. The map was acquired from: http://www.visualcapitalist. com which has many brilliant infographics.

The Inevitable Incompleteness of Intelligence
(Internal and External)

Our identities and our understanding are intertwined

You become what you understand.

Soren Kierkegaard

Recursion and symbols

Because the nature of the "self"[1] is a very intriguing and persistent question, I was really excited to read the book by Douglas Hofstadter called: *I am a strange loop* (his earlier *Gödel, Escher, Bach* was one of the most interesting books I've ever read).

A recurring theme in the book is about the "recursive" nature of self-identity.

A recursive definition of an object occurs when it is defined in terms of itself (or its type). Recursion in mathematics (and coding) is a process in which the execution of a certain function might need to call on that function itself (so the function executes itself to execute itself). Let's look at an example.

The Fibonacci series is defined as: Fib(0)=0 & Fib(1)=1; and Fib(n) = Fib(n − 1) + Fib(n − 2) for all $n>1$. So the series is: 0,1, 1, 2, 3, 5, 8, 13, 21,...

It is recursive because to define the Fibonacci series, you use the Fibonacci series.

But math is easy (?), let's try to apply that to your "self" or "identity": You = ??

We can say that You(now) = You(sometime_past) + Something.

To define the "you" of today, we need to use the definition of the "you" of yesterday.

This is basically saying that previous experiences and knowledge get built into people's identities, and this is a lasting (and even compounding) effect. It is a way of saying your current self = the (weighted) sum of all your previous experiences, but there's an added subtle element. The changes to the self themselves are filtered through the self (the methods, and content, of the different approximations you use are part of that filter).

In fact, we can think of people as being Exponential Moving Averages of their actions [and previous selves], with a varying number of data points, so:

You(Today) = Actions(Today)*a + You(Yesterday)*b

It is recursive, incrementally changing, and subjectively biased toward specific experiences, which are accounted for by the weights ("a" and "b") given to new actions (variability) or the previous self (stability).

We can look further into how this happens. In Hofstadter's book, and in many other places, the self is defined as the owner (and manipulator) of a potentially infinite set of symbols. The symbol is central here. What is a symbol?

A symbol is a sign that points to something else other than itself.[2]

Figure 6: This is not a Pipe (Magritte)

The human mind understands many different symbols and uses them in further knowledge acquisition. Most of our knowledge passes through processes of symbol manipulation. We know a set of symbols, and then we use that set to learn more symbols and relationships, and maybe even to edit our existing knowledge.

Understanding in a sense is a process of extension. We extend our symbol set, and thus extend the self (the owner of the symbol set), but what happens before we can assimilate more knowledge?

It has been argued that a significant part of this is simply a process of analogy, where knowledge and understanding of the new object happens by comparing it (analogy) to already possessed knowledge.

We conceptualize new items in our mind like fuzzy nodes within a network of meanings. We know things by embedding them in this network, sometimes obscuring their boundaries and relating them to different components we're already familiar with.

This understanding is then subjective and fluid.

Understanding is an expanding fuzzy network! So are you.

Two things become evident here. One is that it is very wise for young people to read (learn) more, sooner: "the more you learn, the more you CAN learn." If knowledge happens by a process of comparison to existing knowledge, then if the existing knowledge is richer, the potential is bigger for more learning (assuming processing ability is not a bottleneck). The second, more relevant here, is the hidden approximation in this: As the new object is being learned, and if we assume that the knowledge isn't exact, the knowledge of the object will be something between its actual/true nature, and the (understood/conceptual) object we're comparing it to.

Approximation is central to our process of understanding, but the challenge is to be able to evaluate the error (E) inherent in the difference between the reality of the object and our mental representation. If that is not possible (and it frequently isn't) it just helps to think about the nature of the approximating reference (B), and why we chose to compare to it.

The more active our minds are (and brave), the more we go into areas of the unknown. These are murky areas, and they can be challenging and sometimes uncomfortable. Our minds' expansion into unknown areas can sometimes be iterative too, because we don't go through it linearly, always moving forward. In reality, we go back and forth as we quit the process and restart it at later times, sometimes with similar analogies, and sometimes with relatively fresh ones. With each iteration, we move forward a little, stopping once *some* progress has been achieved.

If you think of your knowledge as a network (or a tree with multiple-connections possible), then a new concept (still unknown) can be thought of as an object near this network (or tree), which isn't connected yet. "Knowing" is about extending the existing network and connecting it to the new object. The nodes connected to the new object determine its perceived (approximated) image in the learner's mind, even if that perception is not the accurate representation.

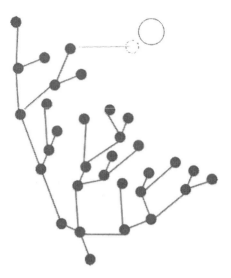

Figure 7: The network of knowledge

The Tree (if you can call that a tree) is our current knowledge. The solid circle is the new object to be known, and it is known (the dotted small circle) as an approximation between the existing knowledge network and its own properties.[3]

An interesting point to note here is related to our sense of self and what truly distinguishes it. If this "self" is a manipulator of a set of symbols, then many of these symbols will be associated with the self (somehow), and will carry emotional weights with them. We will feel positive (or negative) about some symbols, and hold them dearly as they are a part of the "I." As emotional and cultural components get involved, we can understand why some symbols will be used more than others, and why some new knowledge might be distorted as we acquire it because we are trying to make it adapt to our held set of beliefs and "sacred truths" (adjacent nodes).

I am what I see...I see what I am

Some of these points are nicely depicted in an experiment[4] that asked subjects to watch a 45-second video clip of a violent struggle between a police officer and an unarmed civilian. The video itself didn't show clearly if the police officer behaved improperly. This was left to the subjects to decide as part of the experiment.

Naturally, and because what we see as reality is usually created in our minds, the answers of experiment subjects differed. However, the variable that helped predict answers wasn't how logical people were in evaluating the struggle. It was Identification (eerily echoing the result of the batman-ad experiment mentioned at the beginning). The way this effect of identification on judgment happened was telling too.

Before watching the video, the subjects' self-reported identification with the police officers (as a group) was measured and recorded.

As expected, people who identified with the police less strongly (so some statements like "the police represent me," "police officers as a group are close to me," for example, don't apply well to them) were more likely to think that the police officer deserved punishment. Furthermore, this was affected by how often they looked at the police officer in the video (in research, this is called a mediation effect).

The subjects who looked often at the police officer had this clear inclination to ask for a harsher punishment, as opposed to those who didn't. People who made stronger judgment against the police officer typically spent more time looking at him, and made subjective interpretations of the officer's behavior. A similar effect wasn't seen with the in-group, those who identified more closely with the police. This is typical in psychology, where even the most arbitrary of "belonging" criteria can lead to bias and subjective interpretation and judgments.

This police example is a typical case of judgment and actions being tainted by a filtered view of the world. We experience the world not as it is, but according to existing beliefs. Our brain's circuitry has many more interconnections than it has connections to sources of external stimuli. Some scientists[5] suggest that our brain draws an image of the world around us, and uses sensory inputs (from the senses) to adjust and correct that, not the other way around. It is "filling in the details, making sense out of ambiguous sensory input."

The ultimate cause of this imposed filter in our senses in the above experiment, and in many of our daily judgments and interactions, is our image of ourselves and our identity.

This is another experiment: think of a sports fan watching a match in which their team gets into a fight with the opposing team. Whose fault would they think it was (genuinely)?

Judgment (and action later) is shaped by our ongoing experiences, themselves shaped by our own perceptions of our identity, which itself invokes all kinds of affective responses and prejudices. This judgment, often unclear to us, is not objective, but an expression of our ideal or imagined self, and an approximation of its entanglement with the different experiences we encounter!

Leon Festinger (famous for his pioneering work on "Cognitive Dissonance"[6]) says: "people cognize and interpret information to fit what they already believe."

Here the systems' view would be very helpful.

Humans are whole/complete systems of knowledge, emotions, creativity, and linear/causal thinking patterns. Even the immune system's learning is a mix of different general-emotional and causal-linear processes. A system has many interlocked components that affect each other and different feedback loops (recursive), and its current state is a reflection of a history and a vast set of environmental settings.[7]

In this sense, we not only become our experiences, but we also make our experiences of reality in our image (approximately).

This fuzziness of knowledge gained and shared applies to learned experience. Our knowledge gains additional complexity and "coordination" mismatches as we move it into the fields of communications (different people's approximations become part of one bigger – and even more complex – system), and action (bigger sets of unseen variables can enter the equation). Approximation on this individual-identity-understanding level acts as a "seed" for more complex approximations when social interaction is added (through communication and action).

All this complexity helps explain why a simplicity-creating tool is so valuable too. Complexity, systems, and richness will be discussed further in the next (next) section.

The pact with cultural resource sets

Our cultural affiliations serve a vital role that links our identities, the symbols we use, and our environments. "Culture" is a very problematic term, but – for the sake of simplicity and brevity – we will use it here to refer to the set of values, meanings, and symbols that are common to a particular group. A certain culture provides a shared meaning space for a group of people, and is passed on to new members by mechanisms of learning.

We use cultural resources to make sense of the world, and to create representations that connect novel or incomplete information to our existing knowledge network. This generalized definition can be applied to different levels, from daily life, to language, work, the arts, spirituality, and beyond. I believe that our cultural affiliations can't (or shouldn't) be unique or limited, and are best represented as "open sets of resources": Cultural Resource Sets.[8] They are (our cultural identities) pluralistic (so that we can have many different sets of common symbols), dynamic (so they change and evolve with time), content-rich, and context-dependent (they depend on how each individual interprets and expresses them, and on the situation). They come from a variety of sources and origins (so we have national cultures, popular cultures, work cultures, etc..).

The different components of our cultures (resources) routinely carry "excess meaning" that connects them to other elements, values, and assumptions within a particular cultural set that we internalize. In a sense, this excess meaning activates the cultural set, and allows cultural resources to influence our cognitive (reasonable/mental) or affective (emotional) or behavioral (action-oriented) states.

Cultural units (or memes) reproduce, spread, and evolve, in a manner proportional to their utility within our social and cognitive fields...according to how well they help us make sense of (and approximate) the world.

This is why people (mentioned in the prologue) who were positive about American national culture, a certain celebrity, or super heroes thought that they could understand an ambiguous product better.[9] The "radical innovation" in the ad felt somehow more familiar to this group because it was described in terms they loved and were familiar with.

The symbiosis here is that in return for their role in organizing and facilitating our cognitive (and social/practical) lives, we adopt these cultural sets as part of ourselves and perpetuate them.[10]

The approximation powers that they grant us are so valuable to our system of understanding that we give them a part of our identity (and we might completely surrender our identities in extreme cases).

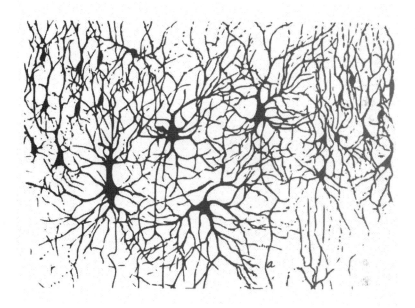

* Seezio's Ship

(Or: It is what it is)

Time passed as Grumis stood there. Watching, recording, learning. The sun, the moon, the stars all moved numerous times (or didn't), and the work went on.

On the face of it, there was so much work done. It couldn't be said, however, that the tower was materializing and getting nearer to completion.

The relation between work and progress was ambiguous at best.

Because the tower was being built without a master plan, it had a peculiar evolution. It was a grand project, occupying a large piece of land. The land beneath it wasn't uniform in terms of its level and constituents. This irregularity of the ground beneath it became visibly built into the tower...

There were places where the ground beneath the tower was curvy, and this had an effect on the structure directly above it. The earth caved in other places, and it was obvious that supportive structures were placed to enforce the building in specific locations, and those – in turn – were used in subsequent construction. The characteristics of the earth and its level reappeared in an augmented manner in the different parts of the tower. A degree of difference on the ground corresponded to 2 degrees on level five, and to 4 degrees on level 10.

Initial conditions caused by the workers were important too. Mistakes had a strange propensity for survival and reproduction. Mistakes that workers made on the lower levels, and that weren't the cause of a break-down or a collapse, were fortified by subsequent levels which were built with more strength and attention, and so ended up being bigger and more sturdy. The result of a worker's mistake was sometimes transformed into a beautiful and large substructure at the higher levels. Any small detail on the ground and the lower levels had astronomical impacts on the higher levels.

The tower remembered its past...and it built on it.

It wasn't just that. New workers were coming and old ones were leaving...all the time. As the new ones came, they did things somehow differently, and the results of their work showed. As you looked from the bottom up, you could see the change in style as if the time axis was transformed into a vertical spatial one.

Time and Space were blending eerily in the tower's rise.

Where groups of workers shared taste and style preferences, they cooperated, and this created big areas that had a similar homogenous feel within, but could be easily contrasted with other similar areas on the tower.

The overall feel and character of the tower kept changing as it absorbed the inputs and efforts of the new-comers. It was open and careless, and no one could say what it truly was.

The tower reminded Grumis of a certain ship of adventurers he had seen. The ship's captain was so fond of it that he didn't want it replaced, no matter how badly damaged it was. He just kept replacing parts, even till the ship was no longer identifiable...After a long time, very few parts remained from the original ship, but to its captain it was still itself.

The effort on the tower was like no other procedure Grumis had seen before, and making conclusions or predictions wasn't easy. Trying to understand it was stressing all his cognitive capabilities.

He had important questions that needed answers...

Forever ignorant: eternal cognitive poverty

*Even our best-tested and best-corroborated scientific theories
are mere conjectures, successful hypotheses, and they are forever
condemned to remain conjectures or hypotheses.*

Karl Popper

The "Rich, Layered, and Out of Control" reality

We live in a highly textured and rich reality, and there is an infinite ocean of knowledge and information floating everywhere around us (a bold claim, I know. But think of all the meta-data floating around). The sheer size of "potential" knowledge, or the knowledge that we "might" try to acquire over a lifetime, has always been fascinating to many philosophers, artists, and scientists. The sometimes-counterintuitive metaphor here is that knowledge is better thought of as a tree (or network, as discussed before) with many branches, rather than a reservoir. The more branches you have, the more branches you can add, and I think that this metaphor is a very simple and elegant way to grasp the exponentially growing size of knowledge.

Not only is knowledge vast in size, it is also highly layered, albeit in a manner that isn't really linear. What I mean by this is that its structure often extends into different levels (hierarchical), where each new successful level includes and absorbs the previous ones, and contains properties that "emerge" from its lower-level constituents.[11]

This – simply put – means that interactions generate outcomes that can't be predicted from units. From a specific perspective (or layer), it might be impossible to view reality as seen by another vantage point (belonging to a different layer).

Think of a human society which has its own characteristics and behaviors (as a unit). A society "does" certain things, while being composed of individuals, each having their own goals and behavioral patterns. Understanding the individual behaviors and psychology doesn't reliably predict the large scale (social) picture. The same (or similar) can be said as we move the unit to individuals, organs, cells, and beyond.

This concept of a layered understanding applies elegantly into a host of different situations. Just as the 5-Whys help uncover root causes of problems by exposing causal relationships and latent reasons, an understanding of the layers on which the different frames of reference of different people can help resolve conflicts and synchronize action and communication.

Think of certain economic problems, like monopolies, corruption, or the underground economy. Discussing the roots and the solutions of these problems can be approached from a purely business perspective, from a political perspective, from a cultural perspective, from an individual perspective, among others. To the dwellers of each of the layers (business leaders, social scientists, psychologists, lobbyists, politicians, etc...), their own priority perspective takes the central stage, and the others are (fully or partially) approximated away or assumed to be subsets of the central (read: their own) one.

These people are all trained to look more clearly through a specific perspective and are most comfortable in projecting the problem into their own plane of expertise – it is more familiar and consistent with a preferred set of values and field of knowledge.

This is human nature. In response to the world's complexity and the nature of the human experience, our attention is constantly shifting between the different layers of reality. We slice reality, and deal with it (usually) one slice at a time. We sometimes keep on moving between these different layers (physiological, emotional, individual, small group, macro, etc...) as we reason and think and communicate, or even work from unsynchronized positions. The different layers correspond to different mental tools and values and notions, and losing track of them can distort the harmony needed for synergies or efficient cooperation.

The different layers and the different perspectives within one particular layer are two dimensions that express the richness of the human experience. The acknowledgment of this takes us deeper into the issue of incomplete perspectives. A result of the complexity and richness of reality is that our knowledge of the world is in a state of perpetual incompleteness. It is not just impossible, but also highly impractical to try to have a "complete" knowledge of a specific area or even situation, because then so much would have to be grasped, making the process of knowledge collection useless and impractical.

We take one perspective within one layer at a time. We prioritize, and this is what leads to action. We chose to make the world smaller as we approximate away layers and perspectives.

Imagine a group of (very diligent) cartographers who set out to create a map so large and exact, so as to accurately depict every detail of a country[12]. How useful do you think that map would be? What is its limit? Which details should it show?

Both the physical and social environment around us are highly complex realities. The word complex, as used in systems thinking, refers to an organization that is characterized by many interacting and interdependent variables, time delays, nonlinearities, and feedback effects.[13] This means that it gets quite hard to perceive and understand the system as a set of static components.

In a sense, and because our knowledge (scientific) and theories are not truths, but mere hypotheses and conjectures that await being refuted (this is what Karl Popper argues beautifully, as in the quote at the start of this chapter), the primary theme of our knowledge quest is progress and explaining more...They are small steps on an endless road – and that is a call for humility.

"My eyes, My eyes!" – make it go away!

To respond to our shortcomings and the complexity, we use all the mental tools we can muster.

We filter away information in different ways, or focus our attention on a specific component of a certain situation, or even use intentionally reductive models to represent reality. This is typically done (consciously and otherwise) to make reality simpler to understand and to represent mentally. We want to make our life easier.

Many of these simplification tools can be thought of as approximations, and this takes us back to the initial claim that thinking in terms of approximations helps reduce the burden of required caution.

Mental models – for example – are simplified knowledge structures that we use to represent the world in our minds (so they are elements on our mental map). We use these mental models to create a simplified (conveniently reduced) mental representation of the world, and this allows us to make sense of our environment and make decisions more easily. Our mental models can impact our perception, information processing, learning, and problem solving [14].

We can think of many of the simplifications and filters of reality that we use unconsciously as approximations and here are some examples:

- By looking at one specific layer of reality, we approximate the other layers away
- By looking at different well-connected systems as things, we approximate interaction (and emergence) away
- By looking at different components of these systems as "things" rather than "processes," [15] because that is simpler, and we're more accustomed to it, we approximate away

many relationships within, and throughout, the different layers of reality

The boundaries of our simplifications often remain completely hidden from us, because we don't invoke them consciously. We grow accustomed to thinking about things in models, and specifically in models that successfully (or optimally) reduce information while still remaining valid at a specific level of observation.

This is the source of some difficulty that results from approximation.

Between ignorance and the fuzzy boundaries of knowledge

In a sense, ignorance simply boils down to incompleteness of knowledge. Ignorance is the state of being beyond the boundary of what is known.

That's perfectly acceptable...if we're careful!

We need this incompleteness to function properly. We can't possibly expect to be able to grasp vast amounts of knowledge and relationships and still maintain proper mental functioning. But as I mentioned, success here is broadly about managing the boundary.

There is a certain place at which we become unaware of all the different filters that are actively reducing the world in our minds, and we step outside the areas in which these filters function properly.

This is because in reality the boundaries are fuzzy, and reality is not made up of solid distinct objects as we'd like to think (because it is easier that way).

The world is not organized neatly in accordance with our (current) ability to categorize it (categories – like things – can be tricky and arbitrary). In reality many of these categories exist in our mind (only). Even seemingly very simple questions like "What is a living thing?" or "What is the color red?" will probably be very different in reality (if there was one) than what we think of them.

Remember that we sometimes approximate according to subjective (individual and group) criteria. Basically, we look at things in a certain way because it coincides with our set of values (which might be different than, or ordered differently when compared to, the values of others), our individual preferences, experiences, or habits...In this sense, it can be expected that different visions of reality will arise, and that some people observing the same phenomenon might see two different things: each of them is approximating different components of the "fuller" picture, and is thus being (intentionally or not) ignorant of them. Just like the projection of the same square from the previous chapter might be a line, rectangle, or square.

Some sad and inevitable outcomes

Internalized approximations become a problem beyond the fuzzy boundary.

When an individual or a group are unaware of the approximation or the information trade-off that they create as they think (or even talk or act), they might assume that this knowledge (or action) is infallible or that it stands for a complete perspective.

When different value systems of different approximators collide, the natural result is conflict.

Approximations – as we've seen – replace something[16] (A) with a more convenient version (B), and we've said that the choice of B (or its construction) depends on subjective factors. These factors can include values that hide below a person's communication and action system. Different people might have different value-priorities, and if hidden behind approximations, they can transform different versions of reality into fertile grounds for conflicts around unclear issues.

Sometimes the punishment is futility and waste.

As we sink deep into approximations of approximations, the amount of complexity that is not properly accounted for in our well-meaning plans renders our efforts futile. We start facing a special category of highly interconnected and "evil" problems that simply can't be solved[17].

Another possible outcome of this cocktail of blind approximations can be the inability to move forward and understand/cooperate. Remember that approximation comes from the word "proxima," meaning close.

Sometimes the enemy of the "best" is not the bad, but the good. Settling for an approximation of the optimal (and possible) solution can result from an incomplete understanding of original compromises.

Pedantic: what is understanding?

How can we define understanding?

It seems that there is significant effort (within "epistemology" – the study of knowledge) to clarify this concept. I don't want to delve deeper into philosophical discussions here, but as you might expect, there are camps.[18] Without being too annoying (I hope), here are some very simplified thoughts.

"Explanationists" tend to define understanding of a phenomenon in terms of the knowledge of sound and correct explanations for it. "Manipulationists" don't really agree, as they think that understanding can be described better by a cognitive agent's (this is a person, usually) ability to manipulate (and thus use) knowledge and representations about a certain phenomenon in other cognitive tasks.

Do you sometimes feel that you lose an understanding of a new concept (or forget it) because you didn't keep it in mind enough? The act of repeating a concept (at different intervals and occasions) leads to understanding as it becomes part of your natural thinking patterns and gets integrated into the accessible network of meanings. The more familiar feels more natural and relevant. The more relevant gets evoked more, and connected to new experiences and knowledge more. It becomes more "central" in the process as it gets connected to more new concepts in one's cognitive map.[19] This ease of recall and the accompanying utility are an essential part of understanding, and they are likely determinants of future experiences.

In this sense, understanding is always a work in progress as new concepts keep getting assimilated all the time.

Kelp[20] proposes that we define understanding (of a concept) in terms of two variables. Maximum Understanding (Max-U) can be defined as the maximum and fully-connected (supported) knowledge about a certain phenomenon. Degree of Understanding (Deg-U) is the level between no knowledge and Max-U that our knowledge reaches.

Our knowledge at a particular stage is always...an approximation of Max-U! The accuracy of this knowledge is determined by Deg-U, varying between 0 and 100%.

An interesting observation here is that because we are looking at understanding as this interconnected and ideal set, and looking at our knowledge as an approximation of this ideal, we can glimpse into the nature of approximation dominating all our cognitive world.

Our knowledge generally advances in a linear and deductive way, as any fan of Aristotle would tell you. The world's complex and non-linear nature will force us to think about disparate things that might extend beyond our "convenient" boundaries (because we don't know enough when we construct them). This naturally makes one apply knowledge from a certain field into another and engage in analogies and approximations.

The breadth of our experience and our attitude might help us here, but it is unlikely that we can avoid the many traps of perpetual ignorance!

Vigor in laziness; strength in weakness

Caution: laziness is a loaded word. Here, though, I use it in the same sense as "low on resources," particularly energy. Also, I – in no way – mean that laziness is bad: some of the smartest and most creative people are lazy by many standards[21].

In this section I want to move from the inevitable incompleteness of knowledge because of complexity to the other side of the coin: the internal limited mental resources that we have at our disposal.

The stingy (unreasonable) brain: preserve processing, memory, and knowledge resources

They are limited and scarce...or at least so the brain likes to think!

I find it amusing that many phenomena in consumer behavior and psychology can be explained (eventually, after asking five – or four – Whys) by the stingy brain's incessant quest to save resources. Be it memory resources, processing resources, or more generally energy resources, the brain's design and actions seem to regularly seek this "conservation" objective.

One common example comes from advertising. The modern individual/consumer is – for example – subjected to thousands of ads every day. In response to this attack, we only give our precious attention to a few of them. We do that because our attention resources are not infinite, and we must carefully allocate them to understand the most relevant (again, a loaded word) elements in our environment. The result: we live in an "attention economy," where people's attention is very valuable and some (many) companies are willing to pay money explicitly for that attention.

Similarly, we don't remember everything. We don't think about everything. Our existing behavioral, psychological, and mental habits do the selection for us, and dedicate our precious cognitive resources to a carefully (not necessarily consciously) selected menu of items.

It is not very hard to see why this is reasonable.

Our psychological and social structures are designed for efficiency because they were (below the surface) shaped by a process of variation (different styles) and survival (selection). The fittest (fastest, most efficient, most energy-saving) survive.

We pay attention to a selected list of things from our environment, and we understand those. We don't have enough energy (and are thus too lazy) for more. But the interesting question is this: what do we pay attention to?

Sticky bias – familiarity as approximator

Individuality has a role here. So one of the basic concepts is that we'll pay attention to things (or stimuli, or ideas) that are "familiar" (in the sense that we know them, or they are similar to other things we know) and thus "relevant" (affect us personally and mean something to us). These objects and concepts are usually – at least to us – easy to understand. Note that the "easy" part means that they are easier to process, and thus there is a positive (virtuous) loop of resource-saving going on here.

This partly explains why some people's perspective of the world is so limited and narrow, and normally passes through the requirements of their daily job or field of study.

These stimuli might be familiar because they are part of our personal experiences and interests, but also maybe because they are part of our value and meaning systems, a.k.a.: our cultural affiliations.

We are biased in this sense – our knowledge and inclinations and predispositions are mostly "sticky."

If we like something, we are – probably – going to (later) like more things that are related to that particular thing, and we continue on that course of increasing specialization (more or less) for the duration of our lives, unless it is consciously broken.

This is the scary cage of sticky bias!

We like novelty, but only as long as it does not depart too much from what is familiar.

In a sense, this is an implementation of the "path of least resistance" rule. Objects in the physical world (and humans!) obey it. The familiar is easier to traverse because it is predictable and closely known.

Think about music (and you can extend that to art in general): we routinely listen to the same music over and over, and we find familiar music more enjoyable, and we even seek music patterns that we already know and love → we have been trained for those. With time, these particular samples become the reference to which we compare music, and we approximate "beauty" in music in terms of the similarity to these familiar patterns [22].

A similar logic can be applied to paintings, films, and stories [23].

This method might also be efficient at first – I use existing knowledge and experiences to judge new ones. But beyond a certain point, one must be willing to experiment further and keep moving...One must consciously seek new knowledge and experiences, that take novel (alien) perspectives into consideration.

The other name for stubbornness here is inefficiency.

This is why it is often advised to actively and consciously seek different music, even if it sounds annoying at first – it will grow on you.

This is why you need to know when an existing system or work arrangement needs to change. Maybe a business needs a redesign. Maybe a certain line of communication no longer works...

The world is complex,
we don't have time/energy to understand it, communicate it, act in it.
We need shortcuts and approximations.
We will use what we already know as the approximation.
We get stuck!

Individual-level limitations: the tragic destiny

The surprise, however, is how arbitrary this is.

An implication of the discussion on familiarity is how much randomness plays a role in determining who we are, and how we judge and approximate things. We – inevitably – have to ask the question: What are these "familiar" things that drive our sticky-learning bias and approximate our world? Where do they come from?

Things that we happen to encounter sooner (because we are affected by our environment, culture, or mere chance) can end up becoming our favorites and the determinants of what things we might like next.[24] They become our favorite approximators (remember Object B?).

This is the scary cage of arbitrary bias!

On a mass scale, this "randomness" component is very creative. Randomness acts as a safeguard against repetition and allows for the variation of experiences on a large-enough scale (big population). It allows for more experiments.

The individual, however, must understand the limitations that come with this model. The model is creative for the collective, but risky for the individual. Many prejudices are simply ridiculous...They are the result of pure chance. Sometimes these "sticky" mental loops need to be transcended. Sometimes existing institutions no longer work. Sometimes an existing scientific paradigm no longer works.

Remaining oblivious and unaware of what these reductive approximations are doing to our mental and behavioral life is a root of many of the problems discussed next.

Sweeping generalizations and a disengaged view

Can you imagine, at this point in our discussion, how attractive and convenient sweeping generalizations are?

Sweeping generalizations are approximations on steroids.

You can just say (or think, or assume) one sentence, and that sentence will save you mental effort related to a whole big group of things (or people). You don't need to think about that whole category of things (or people) again, because you can just attribute any event or fact to this generalization, even if it requires some minor twisting. To you, they won't even need to exist as individual elements, because your "sweeping generalization" takes care of that and puts them all in a conveniently labeled box.

In a sense it is great for saving memory resources because you reduce the amount of things you need to remember, and great for saving processing resources because you reduce the amount of thinking and judging you need to do, because this magic model can encapsulate so much.

This is more likely to happen the less nuanced your interaction with this group is, the more lazy (or resource-cautious, to be diplomatic) you are, and the more filtered information about this group you receive. In a sense, you are not really engaged with this category, and you don't care about learning more about it, so that one disengaging and generalizing abstraction can do.

The potential for this observation is grand – we'll talk about racism in a coming chapter.

Neuro (and other) foundations of passivity: the default-mode network (DMN) and mindlessness

What is your mind doing most of the time?

Try to reflect on that, and try to observe it in the coming few days.[25]

When I started noticing and registering what I find myself (effortlessly) thinking about (cognitive habits), I found out that there were repeated themes that kept playing out. A relatively limited set of related thoughts kept reappearing...And once you notice them, the repetition becomes quite boring, you start changing mental habits, and – as in my case - the amount of self-praise might get reduced.

Later, I found out that this is scientifically well-described, even to its biological foundations. This phenomenon even has certain areas in the brain that are dedicated to it. They are termed "The Default-Mode Network" (DMN).

The default-mode-network, discovered by neurologist Marcus Raichle, spans a number of connected brain regions that show increased activity when a person is not focused on what is happening around them [26]. These areas of the brain work "in default" when we're not really focused, and they play stories from the past, ruminations and reflection, daydreaming, and thinking about the future. They are also related to thinking about stories, empathizing with others, thinking about "our" group, and personal and collective values.[27]

This is very useful for the formation of episodic memories, understanding narratives, and assimilating experiences.

I hope you can see the problem. Approximations are closely related to these default thought patterns, and the more a certain "episode" is played in the brain, the more likely it is to become habitual, and be played even more (familiarity is crucial, remember).

We frame our new experiences in terms of the most mentally present past ones, so a vicious loop becomes likely without attention here.

The more habitual and passive your thinking patterns are, the more likely it is that your thoughts will belong to an arbitrarily (best-case scenario) diminishing set.

This means that you are likely to define new experiences by linking or approximating to this existing "salient" set (a salient idea or set of ideas is one that is easily recalled and mentally present). The same applies when you are more disengaged and not directly concerned with something you're contemplating: it will go through your habitual thinking patterns, where it will be approximated to some outcome of feelings and inclinations.

Where do the thought patterns that repeat themselves come from? In my personal experience this set is quite predictable: previous default thoughts (so there is a recursive/feedback nature here too), cultural artifacts (like books, songs, films, etc...) experienced lately, and mental deliberations around common themes like daily experiences, dominant hopes, and major concerns.

The sticky self and the group (through cultural and social interaction) are filtering our new experiences on unexpected levels!

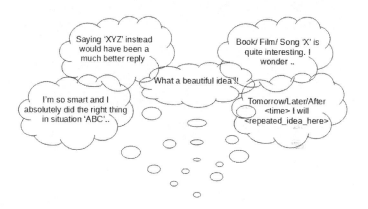

Figure 8: A personal set of go-to thoughts

The danger is this: we are destined to approximate our new experiences in terms of a limited set of personal stories that keep on playing by default in our mind, constantly – if unattended – amplifying our insecurities and anxieties.

Spending time without paying attention, and while being on "autopilot," is in no way something rare or strange. Some studies report that people spend more than 40% of their time "not paying attention"[28], and this state of distractibility has sometimes been correlated with being unhappy. The default-mode network becomes more prominent when people are lonely (expected) and reflecting on their social situation. Depressive people tend to be default-mode-fixated on stories that evoke fear, anxiety, and shame (negative self-talk). It is interesting to note that the default-mode network becomes less prominent with people who practice meditation and can control their ongoing mental chatter. Our neuro-plasticity can help us out of the adverse mental effects of these habits: when we are aware and observe (and thus are mindful of) the constellations of ideas that dominate our idle thoughts.

Again here, taking control seems to be the remedy. Observing and controlling the default mental narrative is a great starting point. Even if it has a brilliant task, our default thinking modes and narratives can – like any self-respecting system with a feedback loop – spiral out of control.

The application of approximation to the discussion of the default-mode network is very illuminating. By summarizing the world, and forcing it through the lens of the narrating self, the links between neurological roots and philosophical and psychological implications become more evident. This requires more introspection and research, of course, but for now I'm content with just introducing it.

Confidence, conceit, conflation

The fundamental cause of the trouble is that in the modern world the stupid are cocksure while the intelligent are full of doubt.

Bertrand Russell[29]

Unknown Unknowns: What if "Ongs" are not "Rongs," and "Bongs" are as Important as "Hongs" and "Kongs"?

Unknown unknowns are extremely important when making decisions in complex environments (and they can be funny too).

An unknown unknown is something that we don't know we don't know about. Let's imagine the following (very plausible) scenario.

To make the correct decision on whether I should buy some "Ongs" I – ideally – should know about the "Hongs," "Kongs," and "Bongs" of the situation. Being an under-achiever in meta decision-making, I don't know that. In fact, I've never heard about "Bongs" at all. I have no clue about "Bongs" altogether, and I especially don't know anything about the "Bongs" of this particular situation.

I will – however – still feel happy and confident to make a decision just based on the "Hongs" and "Kongs" that I know about.

"Bongs" are unknown unknowns: I don't know about them, and I don't know that I don't know about them. I also don't know that knowledge of "Bongs" is essential. It is as if they don't exist, and their exclusion from my knowledge doesn't cause any hesitation or reflection on my part.

Example: I want to buy a unicorn (an Ong). To get a good unicorn, one should consider the mane (Hongs), the tail (Kongs), and the secret tooth (Bongs). I make my decisions just based on the mane and the tail, and have never thought or cared about the secret tooth. I might get good unicorns a few times, but it won't work forever.

I might get satisfactory "Ongs" a few times by making decisions based solely on "Hongs" and "Kongs," but this is just a matter of statistics and randomness now. My (lucky) success would only contribute to my inflated confidence as I remain oblivious to what I really need to know. Worse, by making decisions based on less variables (two instead of three), I might also feel very smart: the problem appears to me easier than it really is.

The particular applications of this (admittedly strange) prototype are too numerous.

It can apply to research, management, public policy, problem solving, statistics, and many more areas. Unlike in the case of a "known unknown" (where you know that there are certain gaps in your knowledge), with an unknown unknown you might feel very confident about your knowledge or action or communication.

In the above case, "Hongs" and "Kongs" are assumed to be a good enough approximation of the needed data to make a decision on "Ongs" (and they indeed might be). But the person doing the approximation has no clue that he/she is doing that as they approximate away the "Bongs" from the decision model.

Another danger of using unknown unknowns can be seen in the inverse situation: adding information. What if "Ongs" to me are "Rongs" (due to certain conditions, since – obviously – I'm not quite sure what they really are), and I don't see any difference between the two, while in reality there is?

Example: Maybe I'm buying a unicorn unaware that it is a separate species, but think of it as a horse (somehow)!

If a certain approximation is not clear, and the person making it believes that they are in fact dealing with the real thing (object A or "Ongs"), while in fact they are dealing with an approximate replacement (object B or "Rongs"), many of the characteristics of B would be transferred to A, resulting in different illusions and conflations. Object B – you see – is not disconnected, but is usually associated with a group of emotions, other actions, and concepts. Its characteristic of strong embeddedness in the behavioral or cognitive network of the person is probably the reason it was used.

This set of associations of object B is then transferred to object A, and it becomes the source of action.

This is what conflation means here: one object is amalgamated with the (sometimes arbitrary) associations of another, leading to many possibilities of confusion, inefficiencies, and conflict.

Next are two examples to illustrate these points from everyday work and management situations.

From the workplace – A

Many work-related stress situations can illustrate the previous analogy.

A certain person's deadline on a major project can be associated with heightened levels of stress and pressure, either because of demands and supervision from levels higher in the organization, horrible bosses, or because of certain perceived risks to the self-image and reputation.

With time, and due to the force of habit, this person will be automatically nudged to treat different projects and work assignments with the same negative attitude and emotional state, even if that means disregarding the new circumstances and requirements.

The ability to balance requirements and resources, and the strategic use of different kinds of resources to achieve objectives are frequently neglected skills. Limited experience and emotional vulnerability will push toward the use of the same frameworks, mental tools, and behaviors in dealing with (somehow) similar problems.

Any new assignment or project (A / "Ong") is treated as if it were that same stressful project (B / "Rong"), leading to a vicious downward spiral of increasing stress and crushed morale, and inability to rationally assess resources and requirements, the calibration of which normally helps with reducing stress and clarifying responsibilities.

From the workplace – B
The situation of unknown unknowns also sometimes applies to managers being promoted from departmental to general responsibilities.
A new general manager with a sales background will tend to see the organizational challenges as relationship ones, and one with a finance or logistics background might be more focused on solving challenges by focusing on processes, etc...
Their previous focus on these fields has worked in the past (confidence), and these are the fields they're most comfortable with (cognitively). From a general management perspective, however, this can have dire circumstances if the newly promoted manager doesn't have an open and flexible mind!

A tale of two illusions: confidence and attribution

So we can have a gap in our knowledge causing approximation, but that's unclear to us. We also have associations that have been transferred to this situation as a result of this approximation. These associations can be values, emotions, or behavioral habits.

Mindless approximations can create two kinds of illusions: false confidence and the mirage of non-existing relationships.

Heightened confidence in one's judgment and skills will result from the perceived level of familiarity and a distorted historical record. Successes can be attributed to personal competence rather than chance, and one of many judgment biases will contribute to this increasing false confidence.

The associations that come with the approximation contribute to the degree of confidence, which might be ironic. The more "connected" a certain item is to our knowledge system or behavioral patterns, the more confident we would be about it (its truthfulness, its correctness, its righteousness).

On the other hand, as we approximate, we attribute characteristics (that might be emotionally charged, or culturally biased) to an object in an inaccurate manner. These attributed characteristics should have been linked only to the approximate (object B), but they are now linked in our mind to the actual object (object A). These are the non-existing relationships, which can add a layer of complexity to our understanding.

If the approximation (A → B) remained an "unknown unknown," however, the confidence in the replacement would result in an assertion that the reduced version (B) is perfectly complete and right, while it may be so only under special circumstances. The error between B and A (E) becomes an unknown unknown in this case too, harder to spot and understand.

Illusions of understanding: hold my concept

As we've mentioned before, understanding is – in a sense – an expansion of the set of things which belong to our knowledge set, by incorporating a new concept and integrating its set of relationships into our knowledge web.

This is not final, though.

This knowledge and newfound understanding is sometimes elusive, and we habitually assume that we understand something, only to discover later that we forgot what it meant precisely if we tried to recall it.

What if I asked about concepts within courses that you've studied in high school or university, but haven't actively used since then? I've tried this and many people are surprised that they can't come up with correct definitions of many (relatively simple and essential) concepts, but rather retain vague links and pointers.

This means that the mind couldn't sustain (hold on to) this knowledge because the new concept wasn't well-integrated within our cognitive network. The concept wasn't held in the mind long enough.

The limitations of our cognitive system can do these acrobatics, and trick us into believing we've learned something, only to discover later that we didn't.

This is how illusions created by overconfidence extend into misjudgments about the depth and breadth of our knowledge. We approximate based on the assumption that deeper down we have a comprehensive understanding, while in fact all we have is a vague familiarity associated with a now-empty slot.

Making good tea is the biggest challenge

I heard this story a long time back, and now I can't even remember where, but there are different variations of it online.

A guy walks into a shop that sells tea. The "tea-guy" (we'll call him that, or "Mr Tea") only makes and sells tea, all day. If he sold 5 cups of tea per minute, it means he'd be selling around 2000 cups per day. The customer walks in, asks for his cup of tea, and as the shop owner prepares it, they have a little chat.

The tea-guy is not happy with many things. He thinks that the prime minister is not doing a good job and is ruining the country (this – usually – is probable). The minister of education is doing a terrible job and students aren't getting the learning they should get. The municipal authorities are horrible, too, as the conditions of the roads and facilities are utterly unacceptable.

The coach of the national football team is an idiot. He called all the wrong players to play for the national team.

The customer recounts the experience, and remembers thinking to himself, "he just stopped short of offering to teach the top football player how to shoot the ball better."

"This tea-guy is really amazing," the customer recalls thinking to himself. He knows more about doing all these different and challenging jobs than the people who've dedicated their lives to them.

The only sad thing is: this tea isn't very good. If only this guy could make better tea!

A more dramatic – and sad – form of this problem was expressed in a newspaper article that had the title: "Why losers have illusions of grandeur."[30] It described the case of McArthur Wheeler.

This man, in 1995, robbed two banks without a mask. Why? He didn't really think he needed a mask, because his face was covered by lemon juice.

Lemon juice works as secret/invisible ink (it does – look up the experiment). That should make his face invisible to surveillance cameras. His misunderstanding of cameras, lemon, and invisible ink (to say the least) led to his fortunate arrest, because his beliefs weren't questioned or tested. He was just too confident (and – surprisingly – was ignorant of his own ignorance).

His is an extreme case – I agree, but this is why it is insightful. Can you imagine someone with moderate knowledge making such stupid mistakes?

Things seem simple from far away. SFFA!

Dunning and Kruger have something to say about this.

"I simplified it, and now I'm sure!" – the dilemma of confidence

The dilemma of confidence is obviously one of the key components of this system of approximation and "acceptable errors."

An inflated confidence prevents us from being sensitive to the scale and scope of approximation, and thus the inherent likelihood of error in our thinking/actions.

Confidence – by the way – is great because it encourages us to act. This is its job description.

It does work in mysterious ways, though, because it is highly subjective and dependent on previous experiences and personal dispositions. Sometimes it can result from repeated random events, and sometimes from bigger psychological complexes.

Confidence is – in a sense – a belief that one's own ability is superior to the requirements of a certain situation. Confidence is a matter of certainty. Again, a very important and crucial trait for success and work in life.

Confidence is about balance (or lack thereof). Caution should be exercised when one's own assessment of a needed set of resources (personal skills, personal knowledge, group functioning...) contains an oversimplification of the difficulties of a certain situation, which – as we saw – is what approximation might sometimes do, especially when left unchecked.

The boundary between simplification and oversimplification is a fuzzy one. But as we approximate we automatically discount certain facts and information, and sometimes our "drive to act" takes the lead, and encourages this tendency to simplify as a method for avoiding further effort (mental or otherwise).

A (dubiously) fun paradox: We simplify the situation and approximate because it is too complex for our current level of competence (!), but the approximation can make us too confident in our abilities. It calls on things we're familiar with and reduces the problem to a compressed version of its former self.

Whereas lower confidence might sometimes lead to more effort and more careful observation and understanding, higher confidence might lead to discounting a certain problem as being simpler than it is, making it easier to miss important facts, and replace them with existing mental representations and wishful models, causing more confidence.

...Like a circle, like a ring...

The Dunning-Kruger effect

The tragic outcomes that result from overconfidence and from the sheer scale of our ignorance about our ignorance (the size of unknowns that are unknown to us) have been scientifically documented in different fields.

People with substantial deficits in their knowledge or expertise will not be able to recognize those deficits. This is the Dunning-Kruger [31] effect and it is a cognitive bias that results from this illusory overconfidence (being smart and competent is better, so there is a benefit in having this self-image or in reflecting it [32]).

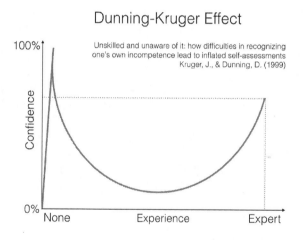

Figure 9: The Dunning-Kruger Effect (1999)

When we know less about something (which is almost all the time for almost everyone), we are more likely to miss our ignorance and knowledge deficiencies. This makes it more likely that we mistake concepts for their approximations or distorted versions, which can translate into more noise in social and communication contexts.

Overconfidence is – more specifically – an oversimplification of reality.

It is a projection of reality onto a distorted, simpler, and more convenient-to-our-inferior-knowledge plane.

The solution to this problem – as Dunning suggests – is to play your own devil's advocate (we're assuming a devil's advocate is more humble than the devil), and question your beliefs and confidence, or to seek more experienced people to cover the blind spots. This might be hard to implement because most of the mistakes and approximations we're talking about here are completely hidden from the subject and happen instantly. We need to learn to live with this bias, especially that we now interact with a wide array of topics in our richer and increasingly complex social and cultural environments (still, I propose – in the final parts of the book – longer-term advice for dealing with these dangers, if you are the cautious type).

Notes

1. The "self" simplified/reduced is the answer to the question: "Who am I?"

2. There are many definitions of a symbol, but I really liked this one: "An instance that typifies a broader pattern or situation," in the American Heritage Dictionary of the English Language (5th edition).

3. I find it very elegant that the birth of interconnections between neural cells resembles what we've just abstractly described as the growth of the knowledge tree.

4. Refer to Garnot, Y., Balcetis, E., Schneider, KE, Tyler T.R. (2014).

5. For example: Lisa Barrett, in an article by Tom Vanderbilt in Nautilus (Issue 19: Illusions): How your brain decides without you.

6. Festinger is famous for his work on cognitive dissonance. Cognitive Dissonance occurs when people experience conflicts between their thoughts, actions, or values. This causes discomfort that people try to avoid.

7. Frijtof Capra (1997) in his *The Web of Life*.

8. Cultural Resource Sets attempt a more practical and somehow stranger – bear with me – description of culture. It is a quantum one, where cultures are seen like spread-out waves with probability functions that can get activated in certain (discrete) circumstances.

9. The broad lines of this discussion are introduced in the research paper referenced before: (Hijazi & Sinha; 2020).

10. For more on this theme, see *Who's Saying Whom?* in the chapter discussing language. The language that we speak is part of the "Cultural Resource Set" that we use.

11. The concept of "Holons" introduced by Arthur Koestler and discussed in detail by Ken Wilber in *Sex, Ecology,*

Spirituality: The Spirit of Evolution (1995) is very relevant here. A Holon is a unit that is complete (whole), but is itself part of a more encompassing whole.

12. Read the short story by Borges, called "On Exactitude in Science" for an elegant depiction of this image.

13. Refer to Gary and Wood (2011)

14. Same reference.

15. One of the key tenets of systems thinking, and one of the findings of quantum mechanics, is that "things" don't really exist as neatly as you'd expect. "Processes" might be more convenient (depending on the context), and "things" can be thought of as "waves of possibilities." Of course, I don't know enough to claim that "processes" are the silver bullet.

16. Here we're assuming that some objective reality exists, but that it is not entirely accepted by many. If we go further, and assume that no objective reality exists, and that people create it with their awareness, then this discussion becomes more evident (and more imaginative!).

17. A more detailed discussion on "Wicked Problems" is provided later when we talk about institutions.

18. Note: The person in a camp usually doesn't believe he/she is in a camp. It is the other guys – only – who are in a camp. Some "objective" observer might see a line of separation between relatively homogenous groups, and – to simplify his own understanding – approximate their positions as two camps.

19. A very relevant extension of this discussion can be found in the "Default-mode network (DMN)" section in the next chapter.

20. See Kelp (2015).

21. The concept of idleness as opposed to "keeping oneself busy" leads to a lively discussion. Check out the essay "In Praise of Idleness" by Bertrand Russell (1932).

22. The "Mere Exposure effect" is a great indicator on what people might like in terms of cultural artifacts and has been documented in different areas. Zajonc is famous for conducting a series of very interesting experiments showing the mere exposure effect leading to preferences of different abstract and verbal elements.

23. The book *Hit Makers* by Derek Thompson has nice insights on the ideas of popularity, familiarity, and newness, especially in cultural markets.

24. I try to approach this metaphorically in the fictional story "The Choice Paradox" (still unpublished).

25. When you remember (or remind yourself), observe the thoughts that spontaneously come to your mind.

26. For more on the Default-Mode-Network (DMN) refer to Raichle et al (2001), and https://www.psychologytoday.com/us/basics/default-mode-network

27. Refer to the article by Andrews-Hanna (2012) for more.

28. Refer to the article by Killingsworth and Gilbert (2010) for more.

29. This quote is from Russel's essay: "The Triumph of Stupidity," which discusses the rise of Nazism.

30. The New York Post (2010): Why Losers have delusions of grandeur? https://nypost.com/2010/05/23/why-losers-have-delusions-of-grandeur/

31. For more on the Dunning-Kruger Effect, refer to the two articles Kruger & Dunning (1999), and Dunning (2011).

32. The opposite bias, "The Imposter Effect," occurs when some people feel they don't deserve their success or status and think less of themselves.

Attempting a (more) complete perspective

The one – The many; the inner – the outer: the AQAL framework

This text is about the impacts of incomplete, invisible, and influential perspectives. It makes sense, then, to use a framework that is particularly designed to give an inquiry a more complete and comprehensive vantage point (I told you (sneakily) at the beginning of the book that frameworks aren't generally bad), trying to avoid some of the pitfalls of the limited view.

Using (even if very superficially) the AQAL framework[1] (All Quadrants, All Levels) can direct our attention to the big picture, and help form a meaningful synthesis (or an attempt of synthesis) of the different areas impacted by approximation in our current discussion.

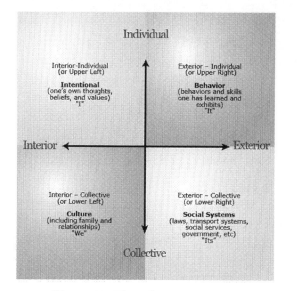

Figure 10: The AQAL Framework

By following the AQAL framework, we can look at the following levels on which approximation can have an impact on our lives:

- Subjective / Individual experiences (Upper-Left quadrant): This field includes internal or self-defined experiences of the individual. These typically span areas of understanding, judgment, and beliefs...
- Objective / Individual variables (Upper-Right quadrant): Here we look at objective, external individual-level experiences, and this can include behavior and other empirically verifiable values.
- Subjective / Group experiences (Lower-Left quadrant): These are the subjective and internal experiences of a collective or a group. This quadrant is about "shared meanings," and culture is typically the central variable of this quadrant.
- Objective / Group variables (Lower-Right quadrant): These are collective objective variables, concerned with the function and structure of social systems, or the interactions of agents within a group. They might include the interactions of organizations and institutions, and even the structures and behaviors of social networks.

Different classes of scientists have focused on one area at the expense of others, thus taking a reduced[2] or approximate perspective of reality. Our examination in the coming chapters will move between behavioral (external-individual) and intent/ understanding (internal-individual), and between social systems (external-collective) and culture (internal-collective). I've tried – as much as possible – to give proportional analysis to the different realms, to paint a clearer picture about approximations and the effects of incompleteness.

Applied: the map of symptoms

The adverse effects of using opaque approximations span all four quadrants mentioned above.

Different approximations traverse personal understanding, judgment, and communication. They also influence individual and collective behavior and institutions, as well as the ways in which groups understand the world and shape it according to their meanings and values.

The apparent symptoms of stealthy approximations include recurring (maybe heated) conflicts where people fight over "hidden variables." These are items or beliefs connected to people's values systems and self-worth, but that have been hidden by an approximation in speech or thought. These values could have been conflated into a situation because of the connotations attached to the objects people use to approximate (B). Crafty politicians are adept at using this aspect to get people excited about something.

Another symptom is the calcification of thought and imprisonment of people in circular and regressive patterns as they get stuck in their own webs of default modes. Reactions and rehearsed patterns of behavior can dominate and limit progress as some approximations take hold and paint an incorrect image of reality. Other symptoms include persistent noise (metaphorical noise, which means communications without highly useful "signals") and waste of resources because of the need to stick to existing (and probably dated) forms of organization and management.

All this might seem a bit too complex, but in the coming chapters I'll try to look at some of these fields and show the symptoms associated with them. A more detailed explanation and many examples should clarify the bigger picture.

From individual decisions and how they are affected by heuristics and shortcuts, to how science, scientific progress, and the social promise of science might be stifled by the approximated view, to the different ways in which language shapes and is shaped by approximation, a deeper understanding of the performance of individuals, groups, and systems under uncertainty can emerge. We will also look at different institutions (political and managerial) and how marketing and brands fit into this framework.

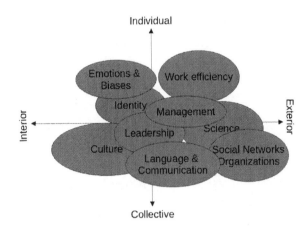

Figure 11: Different disciplines and their AQAL-grid-locations; Sizes and Locations are (very) approximate

Making messy decisions: heuristics, biases, and greatness

(Approximately) rational decisions

How are rational decisions made?

The answer is quite simple.

Rational Agents (people or robots or things or groups) make decisions by following a clear set of steps. They are – after all – rational, and this usually means that they want to maximize their utility (or happiness)[3] as an economist might tell you (extreme caution recommended because of high concentrations of all the symptoms of the Inevitable Incompleteness of Intelligence with this group).

To make rational decisions, agents follow (more or less) the following set of steps. They should (any basic textbook on decision-making should have some variation of this list):

1. Understand the environment and create a comprehensive list of action alternatives,
2. Identify and evaluate all relevant pieces of information (cues) associated with each alternative (like the "Hongs," "Kongs," and "Bongs" of the previous chapter),
3. Weigh the value of these cues,
4. Compare the value gains (against specific objectives, and considering the cost-benefits) that taking each alternative will generate for the decision-maker considering their objective,
5. Select the one that maximizes this value gain.

Excellent...

If it weren't for the minor problem that all this is completely imaginary and impossible.

Unfortunately, we're not designed for that (by now I hope, dear reader, that this statement is not much of a surprise).

The process above is rational, but is highly impractical and doesn't apply to most of the decisions people make (even/ especially managers and leaders). The reason for this is mainly (but not completely) effort.[4]

Remember in the first chapter when people assumed they understood radical innovations better because they liked the statue of liberty?

It is just too much hard work to follow the rational recipe for ranking choices, gathering the needed information, weighing it, and comparing alternatives. This puts stress on our cognitive abilities in terms of the resources needed to process, store, and recall information.

The limitations on our cognitive resources (the energy of our minds – metaphorically speaking) are the/a reason we are not rational. People generally operate (I think at best) under conditions of "bounded rationality," the term coined by Herbert Simon who argued against the adequacy of some models within neoclassical economics.

As we've seen in the previous units, there are many different categories of difficulties associated with each of the steps of the rational decision-making process mentioned above (which we don't need to discuss in detail here), with the environment's complexity, and with the many layers of reality, and unknown unknowns that lurk in the shadows.

Beyond effort and limitations on cognitive resources, rational processing also requires too much time (granted that time and effort can be thought of as interchangeable, but not always), and emotional control. After all, can we confirm that all people want to be rational. Or happy? (revisit what Nietzsche said).

We want to be rational, but not to the extent of causing ourselves discomfort.

At the very least, we want to feel as if we're rational (because that's what good rational people do), or we want to give the impression of being rational (because we want other good and rational people to think that we – too – are good and rational people).

In other words, "approximately rational" seems good enough...So, we use shortcuts!

Heuristics and biases: our convenient daily dose of shortcuts

We don't really need to be rational. We "satisfice" instead. That's just a funny word for "do our best."

We use heuristics, which are mental shortcuts or rules of thumb, to arrive at decisions that are good enough, without bothering ourselves too much. Heuristics are "methods for arriving at satisfactory solutions with modest amounts of computation."[5]

Heuristics are a specific type of approximation. With heuristics, difficult problems or decisions are approximated as simpler ones. Sometimes decision variables are replaced with other variables. This could be because the new ones are easier to process, or are more prominent in our thinking (remember geometric projections?).

Some Heuristics we use to avoid having to think too much include using brand names, relying more on information we receive first (anchoring), country-of-origin for judging products, endorsements, judgment by the best (or least) attractive aspect, likability, scarcity, and others...

These heuristics try to reduce our processing load or the amount of information (or time / difficulty) needed to make a decision. They reduce the number of alternatives we need to examine, or the cues associated with the different alternatives, thus approximating the decision process according to systematic criteria (which include previous experience, prior knowledge, and even personal preferences and subjective evaluations).

Heuristics are useful, but can be dangerous because they can frequently lead to unforeseen problems. Sometimes their use removes critical information. They lead to a significant set of "cognitive biases," which are systematic errors in thinking. We say "systematic" errors because they are not arbitrary but happen according to certain predictable and repeating patterns (remember that these are – generally – a more predictable kind of error). This happens because of the way shortcuts (heuristics) attempt to approximate and simplify the world, when decision makers are unaware of some of their inclinations.

There are many well-known and quite famous biases that cloud judgment. There are different classifications of them, and I've seen very interesting infographics that show the list (which reached up to 188 different biases on some online sources[6]) of biases, sometimes classified into "families of biases."

Here are some interesting and archetypal biases which have been repeatedly documented in research. They also relate to our discussion, and I've introduced some aspects of them in the previous chapters:

- Confirmation Bias: People prefer information that confirms their existing beliefs and values, and validates their sense of identity. They tend to search for, prefer, and recall this kind of information much better than information that is contrary to what they already know/like. They also might interpret neutral or uncertain information as supporting their existing attitudes.
- Egocentric Bias: As part of identity preservation, people tend to view themselves in a more positive light than reality. They also tend to rely too heavily on their own (naturally limited) perspective. Most people (~ 65%) think they are smarter than the average person[7], and a better driver than the average driver. Three studies have documented an ironic bias in which people are biased

to see themselves as less prone to bias than others[8]. This means that our own unknown unknowns are more unknown to us than the unknowns of others (!).

- Attribution Bias: This kind of bias is a set of systematic mistakes that people make when trying to judge the causes of things. We – for example – tend to attribute our successes to our internal qualities, and our failures to the environment. We also tend to attribute other people's mistakes more to personal factors than situational (environment) factors. We sometimes interpret other people's behavior to be more hostile than intended, and we might – because of conflated variables – attribute results to the wrong reasons as we confuse correlations with causation relationships.

- Availability Bias: Availability bias means that people tend to rely on examples that come easily to mind to make judgments and evaluations. This bias might lead to overestimation of the value or prevalence of a certain specific issue or phenomenon because a decision-maker happens to be close to it. This kind of bias is the reason people – for example – fear (unreasonably) homicides or plane crashes more than car accidents. A few media reports about a certain problem make it more "available" and can make people overestimate its importance (many of these biases are good friends of professional manipulators as you can see), and at the same time prefer to read/know more about it because of familiarity (leading to more media reports leading to more availability…you see where this is going).

In all these kinds of biases, there is a systematic reason the error is made over and over. It is usually about easily available (or close) information, and familiarity. Even if the goal is to preserve cognitive resources, the bounds of rationality have to do with identity, previous knowledge and experiences, culture, and even emotions.

The truth is what I feel: the disproportionate power of the irrational

As we make decisions, even those we feel are rational and reasonable, they get twisted and distorted by numerous mental shortcuts that we take. These shortcuts cause systematic biases.

Our own emotions and affective states have a great role to play in generating these biases. The effects of emotions get more intense the more a certain line of thought or judgment is seen to represent what we define as our values, our personality, and our sense of self.

If things threaten to weaken our sense of self, or disrupt the significance of our values undermining truths that we hold dear, we might find ourselves avoiding certain facts, changing our behavior, or twisting interpretations to restore the status quo.

We crave the stability of our rooted cognitive structures, and are willing to go to great lengths to avoid the effort needed to explore and absorb new beliefs!

Nobody wants to open up their inner safe fortress. They need to, however, in order to grow. We live on this tension between stability and progress.

An attack on the "self," for example, or a threat to a value system can be seen as a cause for defensiveness and anger, while self-assurance can lead to positive resonance.

Our self-image, or our concept of our identity, mobilizes our feelings and different cognitive processes to protect itself against influences. It – like almost everything else – seeks survival...sometimes at the expense of our own growth and acceptance of the variety in the world.

On the other side, even mundane moods or temporary emotions can cause distortions in our rational decision-making ability because they "color" the world. Our emotional states are associated (on the physiological level) with specific hormones that – literally – flow with our blood. We don't have separate rational and emotional selves – they blend to produce outcomes.

Since we often rationalize decisions after they've been taken (so a decision is made behind the curtains, and we then seek possible rational explanations), one can think about who – really – pulls the strings.

How are all these emotions giving rise to approximations that shape our decision-making and our view of the world? How is our existing self-image distorting our evaluation of the world and our interactions with it to preserve itself?

On sports and nations: I am (almost) great!

A great example that can teach us a lot about biases, identities, and approximations is the tricky situation of belonging to a group. Nationalism and fanatically supporting a sports team are not the only cases, but they provide amazing insights.

The psychologist Henry Tajfel (in his 2004 research) finds that assigning people to groups, even if done on a completely arbitrary (and weird) basis, like dividing people based on how they interpret abstract art or count group of dots, might lead to certain types of bias for our own group (in-group) and against others (out-group). Cognitive scholars believe that cheering our favorite teams stems from the same subconscious motive that makes us patriotic: we have a clear tendency – maybe with evolutionary roots – to belong to a group!

Note that this is no simple or shallow matter. Some studies[9] find that the level of some hormones (associated with achievement) might get elevated in fans whose teams win, and other hormones might have similar fluctuations between players and fans.

So why this bias?

There are many theories.

Groups are comforting. Cialdini talks about wanting to 'Bask in reflected glory' (BIRG), where we can attribute some group success or bigger values to ourselves. We like grand narratives, and we love heroes and the epic struggle of life and death. We don't like to be small, and groups provide common stories and symbols that can unify us and give us a taste of grandeur and immortality. Interestingly, fans of sports teams (and group members more generally) report a higher sense of meaning in life, and they score high on a number of well-being measures (like self-worth, feeling connected, positive emotions, energy, etc..).[10] The need to belong to a group seems to grow when the self-image gets shaken.

We love our groups (team, country, tribe...) for all these reasons, and for these same reasons we can sacrifice common sense and accept taking quite biased positions on many issues to protect our "ego" and to preserve for ourselves the possibility of getting glory without needing to personally put forth the required effort and sacrifices for it.

Typical results of these biases include – on the extreme – overlooking clear evidence and rejecting perfectly common arguments[11] (sacrificing reason and using your brain less is a small price to pay for greatness).

The group's glory is (approximately) my glory! The group is approximately me!

(Belonging is nice, but get your own glory, and belong to a group consciously and purposefully! Don't be a full-time Birger.)

Special case: Angry Racist Idiots (ARIs)

The best lack all conviction, while the worst are full of
passionate intensity.

W.B. Yeats[12]

I want to take the previous points a bit further and talk about the special class of the "Angry Racist Idiots," or – more endearingly – ARIs. This is a group that really takes biased reasoning and heuristics to its frightening limits.

First, racism is one specific kind of bigotry or fanatic belonging to a specific group as described in the previous pages, where the in-group and the out-group happen to be racially determined (as we saw any arbitrary division might work similarly).

In a very simplified and simple form, racism starts with the need to make sense of the world's (the social world in this case) complexity. It is very hard to understand the granularity involved with people's individual personalities, and their rich personal, social, and cultural experiences – it is just too much effort. If only we can put a bunch of them (people) into one box, label it, and just simplify our work.

Let's approximate THIS group of people into a clear racial "them." That's convenient and easy!

The lazy, evil, unscrupulous, and treacherous "them" are against the hard-working, pure, and intelligent "us" (one of "us," after all, invented some widget (maybe) 732 years ago, and it is very sane and logical to conclude that I must be great too, just like that guy, since we have similar shades of skin color).

I can "BIRG" (Bask In Reflected Glory) and reduce the need to think in one sweeping move.

Bias, discrimination, and prejudice are wagons in the same train. This is what the "Angry Racist Idiot" thinks (or more precisely, doesn't think).

I can approximate this bunch of people into this group (one problem is easier than many).

I can approximate this group into (only) the carrier of these negative traits that fit nicely together and are easier to remember (lazy, evil, dumb...).

I can approximate myself into a member of this other (better) group.

Win-Win-Win!

Easy and convenient feel-good for the mentally lazy.

"Projection" from the early chapters fits well here too. ARIs are always projecting things into spaces with less and less dimensions so as to fit into the limited mental bandwidth of their understanding.

Note: The Halo effect, which is a bias defined by Thorndike (1920), helps illustrate this point from the flip side. We assume that good-looking, well-groomed, and polite people are good, or that any new product made by Apple® is well designed, because we extend the positive evaluation from one dimension to the next.

But why are they "Angry" then? You'd think that convenience was supposed to make people calm and content (even blissful)?

Well, this is not the whole story.

My "good" group has lofty values and ideals.
We stand for what is right and good and beautiful.
We stand for what is "godly" (God likes "us" the most, obviously).
It is very unfair what these evil others are doing.
If I can't get angry to defend the most supreme of values, and to stand up to the grave injustice, then life has no meaning.

That being said, and because of the way our societies are structured, and because of trends, many ARIs are now HARIs ("Hidden Angry Racist Idiots"). HARIs veil their anger and their prejudices and hate, but you can always "approximately feel it" through their language and attitudes overall by comparing to the above reasoning methods, including the superior in-group, the righteous and godly and often-repressed "us," even though "we" were obviously greater once.

Unseen approximations are everywhere, and they're behind some of the most dangerous stupidities imaginable.

I (approximately) want this! Approximation in marketing and brands

The more marketing and branding cases I examined, the more surprised I was by the extent to which approximation, among the other themes discussed here, is prevalent in communication and understanding. In the coming pages, I'll briefly describe a few market-related aspects where incomplete knowledge and approximate conceptualizations are used to make decisions, communicate ideas, understand arguments, and drive emotional reactions.

I initially had this chapter included as part of the previous one, but later decided to include it separately because of the scope of phenomena discussed here, and their substantial social, economic, and cultural implications. There have been many studies that specifically categorize different cognitive biases related to marketing elements like brands, symbols, and endorsements, and we'll rely on some of these in the upcoming discussion.

Frictions and mental gaps: fooled by the market (and the lazy brain)

Why do people overlook important – even readily available – information and make inferior purchase decisions (meaning: you buy something that gives you the same utility for a higher price compared to something else, or a variation of that)? Choosing options that aren't the best fit seems to be a widespread phenomenon that includes different sectors.[13]

The following cases have all been documented in different studies:

- People buy branded drugs instead of equivalent but less-expensive generics, even if the information is clearly printed on the packaging

- People seem to consistently choose the wrong cell phone plans given their previous usage patterns
- Consumers exhibit substantial inertia in choosing their retirement investments (which basically means that they'll stick to whatever is already there), and frequently pay premia for higher-fee index funds

Keep in mind that the above examples (and many more) fall specifically within an area that should (typically) be rational decision-making. You'd expect people to be the most rational with these decisions in particular. The criteria surrounding them (simple comparisons, small number of performance attributes...) make it unnecessary to conduct a deep analysis of other influences like emotional drives, lifestyle, aspirations, experiences, etc...Still, even with these relatively "good" conditions, people don't really behave rationally.

Even in situations where the expectation is rationality, rationality does not prevail.

The research on the possible reason for this seems to fall into two broad categories: *frictions* and *mental gaps*.

Frictions refer to the costs of acquiring and processing information. This is the "laziness" argument we introduced as we talked about the brain's incessant drive to conserve resources. Mental Gaps refer to the use of the wrong mental models or mis-evaluating the importance of specific (missing) information. In other words, this is (roughly) equivalent to the "unknown unknowns" factor contributing to our "inevitability of incompleteness" hypothesis.

Familiarity, inertia, partial blindness, and feedback (persistence of the past) are among the factors that make consumers reject clear and available information within (relatively) simple selection criteria.

Until consumers become much more rational (Can they ever? Should they?), or information becomes more easy to transmit, process, and store, brands will remain super heroes of markets!

Your branded brain

What are brands really? Is it the name of the product? The Logo on the package?

Generally, a brand is defined as something "shaped by the company, owned by the customer." The brand is really located in the customer's mind, and it is the connector (nexus) of different stories, images, sounds, smells, tastes, feelings, and even places associated with a certain product, experience, or service. A brand is the sum of all the mental connections and experiences people have around a product (or a set of products), and it is a value system.[14]

Brands have a very evasive nature and are hard to pinpoint exactly, but their effects are not hard to see.

Brands work because they make life easier, and they make it easier for buyers to understand products, make comparisons, and decide. They – crucially – also provide a structure of stories that we can fit ourselves into.

Remember the rational / additive decision model? With brands, many of the alternatives' cues are replaced (approximated) by a simpler set of convenient visuals, catchy slogans, stories, and memories.

People are willing to pay more for branded products, even if they know that the alternative is probably the same[15]. People trust brands more, see them as if they were human and think of brands as if they have personalities.[16] People feel good about the "familiarity" component contained in these "friends" and acquaintances.

That, by the way, doesn't stop at brands. Think about companies in general: we say "Nike did this," "Apple created that," "Google supports X," etc...In reality, though, this is all an approximation.

Companies don't "do" or "believe." Companies are a group of people, and usually not democratically organized (so "companies" are really a small group of people, or one person). Clarity sometimes is easily achievable and highly valuable.

We – all too easily – anthropomorphize brands, companies, organizations, and even countries. This is sometimes more convenient, more (less) politically correct, less mentally demanding, and overall just easier. It also makes it easier for responsibility – when/if there is foul play – to get dissolved into a soup of indeterminate subjects.

The approximation of a system (brand or company as a network of symbols, actors, and processes) into a thing (brand as a semi-human being) makes life easier. Buyers are happy to approximate the journey of looking for specs and understanding the products that they're buying, and the policies they're validating, with the image of a brand in their minds, or with an image of themselves within the brand's ecosystem of stories and values.

Images are more powerful than words, and can be processed faster and easier.

It is a Win-Win again...

Or is it?

The issue is not that simple. Interesting research is showing that brands have effects that extend beyond what most consumers understand. Brands are used as proxies for our own personalities (my personality = the brand's personality), and consumers might be thus driven to buy products that they think will affect their own personal image (social status, for example) even if they don't feel they need them.

Brands are extensions of people...!

Even more interesting, research has shown that people's actual traits vary depending on the brand they're interacting with. People who were primed with the "Apple" brand (this roughly means that they've been shown the Apple brand and made to think about it), for example, were found to be more creative (better performance on a standard creativity-measurement test) than those primed with the "IBM" brand.[17]

Brands have the potential for deep influence, not just on our choices. They can – at least temporarily – change our personality traits as we respond to the mental images that they evoke in our minds. Our proxies for decision-making, end up changing who we are!

People are extensions of brands..!

Think about that as a recursive feedback loop, and the ever-present role of identity in approximation!

Symbols in ads: countries, brands, and celebrities

Brands are just one category of symbols that are frequently used in marketing and advertising. A symbol is an item that represents something other than itself (it points to something else by resemblance, convention, or association). We are – as cognitive beings – manipulators of an unlimited (expandable) set of symbols (recall the discussion in the chapter on the nature of identities). Symbols are commonly used in advertising to convey meanings efficiently and quickly.

Ad effects are sometimes analyzed by thinking about semiotic triangles. A typical semiotic triangle in an ad happens when a product is associated with a specific symbol, and then the meanings associated with that symbol are connected or transferred (in the viewer's mind) to the product itself.

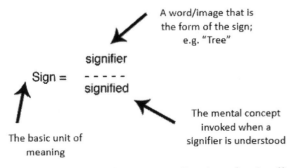

Figure 12: Saussure's conceptualization of a sign [18]

Countries as symbols

General symbols can be used in ads, but also specific types of symbols like countries (yes, again!) and celebrities are pretty popular. A typical and overused example here is Häagen-Dazs. Why does an American ice-cream brand (from the Bronx) have that name? Because it sounds northern-European. And why is that good? The symbol's properties (exotic, expert in sweets, craftsmanship) are transferred to the product (more precisely to how consumers perceive the product). Why does the Starbucks guy tell you "Grande or Venti"? Why is the Audi slogan like that...?

"Country-of-Origin" is used to achieve a specific effect in advertising. Something made in Italy is assumed to be fashionable, something made in Germany is assumed to be well-engineered, something made in Japan reliable, something made in Denmark well designed, and so on...It is easy to remember this. We approximate the country as a set of abstractions, and then we approximate the relationship between the product and the country along that specific dimension.

Symbols are omnipresent in ads, and these include items related to family happiness, health, beauty, wealth, and many more.

Celebrity and expert endorsements

Celebrities are another type of these symbols. Celebrities are associated in people's minds with success, wealth, and popularity. Their endorsements of products are done in such a way that when the customer later thinks about the product, they think about the celebrity and what that celebrity means. This – again – is a case where the product's rational evaluation is replaced by a vague approximation that relies on the assigned celebrity's fame and reputation.

This helps people make decisions, but as you understand what's happening, you might find yourself annoyed at how irrational this can get[19].

A more general case is using experts to transfer their "authority" and "power" to the product. People seem – for example – to believe that anything they're told by someone wearing a white coat is more credible (there is an actual experiment on that). The qualities of competence and confidence frequently flow from dentists, engineers, and scientists to things like toothpaste, machinery, and snacks[20].

Who's choosing whom: the market's many faces

The previous points are an excellent application of how the inevitable incompleteness shifts perceptions and actions in the market.

The methods in which we approximate our purchase decisions (and market-related actions in general, like evaluation, usage, disposal, and promotion of different products) reflect how our identities are dynamically and continuously constructed. Our character or self-image affects – and is affected by – how we approach the market, and by our struggle to represent a highly complex reality incompletely by using limited tools.

The outcomes of these interactions are a function of our sensitivity to the overall process, and to paying attention to our thoughts.

Every consumer (theoretically) has the power to choose, and that very choice, when aggregated, generates a powerful market force. Buying decisions drive trends and inform business leaders on what is acceptable and what isn't. Conversely, we can say that brands also choose their users and shape their imaginations, knowledge, and experiences. The interactive symbiosis is a typical relationship of a recursive and self-adjusting system.

This very complexity remains hidden from most, and part of the system's potential gets clouded in the many approximations happening all the time.

Products are approximated as stories, characteristics are approximated as countries and people, people and groups are approximated as fictional beings, to name a few. All this adds certain elegance to the system, and enriches the set of human experiences. It also throws many inefficiencies (good or bad) into the pot, with effects on our psychological states, well-being, and social and cultural lives.

A wider understanding of some of these processes will be great for common well-being, and is already happening. Technology facilitates that, but one mustn't rush to the Techno-Utopian conclusion that the incompleteness can be easily bridged by a set of new shiny tools.

Social media and information technology advances do help with providing easier access and flow of information, as well as with creating new and parallel communication channels. However, the core social and cultural processes are just that: social and cultural. They will interact with technology, but don't assume that deep and sustainable change to the "governing dynamics" controlling the markets is a technical matter – it rarely is.

People with silicon rose-colored glasses will argue otherwise.

Think carefully about your position in the recursive self-adjusting loop of the market. Think about how your decisions and responsible choices shape brands, and about how the market is slowly shaping your thinking and defining your identity... one approximation after another.

What will (can) you do about it?

* Tawsfus of Segrob

(Or: The price of their souls)

Grumis noticed doors in the ground near the tower construction site. A man would come out periodically, and he seemed ill-tempered. Sometimes he would shout and fight with his colleagues, and sometimes he would approach the construction area to give Zif different tools.

"Who is that?" Grumis asked once after he had left.

"Oh – that's Tawsfus...He is the chief knower. He's helpful sometimes."

"Is he the leader of these other people there...?"

"Not really. The knowers are a peculiar bunch. There are different groups, but Tawsfus is the most well-read, and so he has the most authority. They all are quarrelsome without being physically dangerous, and get irritated easily."

"The knowers...what do they do?"

"Well, they spend their time learning. Very noble if you think about it. They dwell in a giant structure that spans miles and miles underground. It is called the Segrob library. There is so much knowledge recorded in the different books in the different rooms, and the knowers dedicate their life to learning that and producing more knowledge."

"It must be helpful to have them around then?"

"I know what you're thinking. You asked me before if I ever tried planning our work better, and understanding how to go about our tasks. Why don't I ask them for help?"

"Exactly! Why?" Grumis said.

"Well, I actually tried before. Many times, with many predecessors of Tawsfus, and even with him. It was of no real use. You'd think with their knowledge much more can be achieved..."

"What happened?"

"Well, it seems to me they always rush to make claims and assertions, only to be surprised later that this isn't how things really work. Also, whenever a new chief knower attained that position, he would cancel the work of his predecessor, and focus us on a new – and sometimes different – direction. I later saw why..."

"What do you mean?"

"The Segrob library is really grand. I haven't been there, but I've learned about it from many stories. Rooms and rooms of endless shelves. The knowers don't go around a lot. They usually get comfortable in one place. This is where the problems start...you see the books in the library are arranged by topic!"

"Why is that the problem?"

"What matters the most to them is where they start, and what books they happen to read. That becomes everything they care about. That becomes how they see the world. Through peculiarities and details unknown to most of the others...even their own kind. Because they learn a lot, they think that the others don't understand things they do. The problem with what they know, however, is precisely that: that they know it. If they know it, it is because it is a limited perspective. What he tells me is a summary of the summary he remembers of the summary of the books that happened to be in that one area of Segrob!"

Grumis was trying to understand, as Zif went on...

"There was once a chief knower who tried to do something differently...His name was Wawsmutz. He was picky about books he read, but his advice was still strange and impractical. He went crazy later. Some of the books are absolutely inspirational, but I think that some of them are completely unreasonable. I'm not a knower, though, so you can't trust me on that."

"But you could see if their advice worked...?"

"Well...it takes too much time to see the results, and they change too quickly. The knowers' advice was never really helpful to us. We found – with time – that it was better to let them work on their own, and whenever they gave us something we can think about whether it was useful to us or not."

"So that's why planning and knowing in advance wasn't a solution."

"...and that's why what we do is build the tower!"

Zif continued moving the rocks with a grin...To Grumis, there was little difference between Zif and the people of Segrob, each going on with their infinite tasks...to Zif, however, there was all the difference in the world.

The fuzzy truth of science: linearity, reduction, and belonging

*Science is the ultimate and infallible tool of knowledge and progress.
It will save mankind from the different problems it faces.
Scientists are smarter, and they know exactly what they're doing. We
should trust them to make decisions for us.*

Science and the scientific method are surely among mankind's
greatest achievements and tools, but the above statements could
be seriously (dangerously) flawed.

In the coming few pages I encourage you to think about science
as a social force and reflect on its cultural and social role today. I will
try to show how incomplete understanding, overlapping concepts,
and unseen approximations make us (society, communities) view
science (and scientists) in a specific and inaccurate way. I will also
discuss how scientists (as a community) mis-evaluate science's
potential and position, as they themselves are not immune to the
sinister effects of numerous (and multi-layered) approximations.

About science and how it happens

Science is generally thought of as the organized effort to explain
and predict. It is a systematic approach to knowing. The sciences
are seen to belong to one of three groups: (1) Natural Sciences
(physics, biology, chemistry...), (2) Social Sciences (psychology,
social sciences...), and (3) Formal Sciences (mathematics,
computer science...). Each can be further considered to have two
types: theoretical/basic and applied[21].

That being said, there are many discussions on what
constitutes real science, and what method science should follow
to make its conclusions. The jury is still out on that.

Is induction (going from a limited number of observations
to a general rule) really valid (trustworthy)? I've seen no black

swans, and I've looked for years – is it a valid conclusion that they don't exist? Is knowledge extracted from vast data valid (data science)? Is logical deduction valid, with all its abstractions (making conclusions based on premises and logical relationships)? Are simulations valid knowledge tools? What about using analogies?

I will spare the reader many -isms here[22], but it is helpful to remember that the validity of our knowledge and the reliability of our interpretations are not simple issues; they are – in fact – highly contested. These issues are far from settled and clear, especially to scientists (as a species, not as members of groups).

How science happens: le paradigm est mort, vive le paradigm!

There is one theory of relativity, formulated by one of humanity's most brilliant minds. Something of that scale requires extraordinary imagination (Einstein – justifiably – thinks that imagination is superior to knowledge), true, but it, more importantly, requires extraordinary circumstances. It is not simply an individual achievement.

Science is a continuous process, and the bulk of the effort to move forward – contrary to what most people think – is not heroic acts of revolution. It is usually the exact opposite. Large groups of people work on really small (and frequently insignificant to the average person) problems. This is – by far – the biggest portion of scientific work. The dominant method in science (Hypotheses → Deduction ; Verification) is by nature linear/sequential and slow.

Climbing or crawling are more descriptive of scientific progress than flying (or even jumping).

And this is neither an insult nor is it a problem...in itself.

This is because most science work is "normal" science that happens within a specific "paradigm." A paradigm is a "logically consistent" perspective of the world, that is also consistent with the different observations that result from this framing.[23]

A scientific paradigm is the currently approved approximation!

Revolutionary science happens when one paradigm is replaced by another (paradigm shift), and this typically happens when a certain set of unanswered questions according to the previous paradigm become more numerous or more important.

The catch is: the replacement of a paradigm is not a purely logical and scientific process – it is a social process. Scientists (or the relevant authority) guard the existing paradigm by their work and power (and by their very identities), and they decide if the stage when the paradigm needs to be replaced has been reached.

This is why the theory of relativity had to happen in the right circumstances.

The Newtonian model of the universe was too old by then, and technology and research had advanced enough, Riemann's work in geometry was crucial[24], and the community of scientists punching holes in the wall of the dominant physics paradigm was too big.

New science tries to push the boundary of knowledge (at the edge of the paradigm) further, as much as possible. Sometimes the existing paradigm crashes, and a new one (with a new wall) is constructed.

When scientists work within a paradigm, they are working within a limited view of the world that accommodates most of the observations done till that stage. A paradigm is an incomplete model – a simplified and approximated version of reality, or a projection onto a plane that has a few variables approximated away.

In a sense, scientists are – most of the time – guardians of the current approximation of reality. The current approximation is just that – an approximation. It has its mistakes and incompleteness, and its own removed dimensions.

Science is only human
Missing the sweet middle

Science – at any certain stage – doesn't have a "truth" (in the sense of objective and ultimate truth), but rather has an explanation or interpretation that – so far – hasn't been falsified (proven to be false). In fact, one of the most important tools of scientific inquiry is "falsification" and philosophers of science (most notably Karl Popper in different talks and publications) argue that this is the primary way in which scientific progress can be done. All scientific theories have to be falsifiable, in the sense that they should be open to being examined and falsified. Falsifiability is superior to verifiability in this sense.

Science seeks the truth, not certainty. This is a very important distinction. If it sought certainty, it would compromise the falsifiability dimension. Science is humble and a "work in progress," and the results of scientific research normally come in the form of a statement with a certain degree of confidence (95%, 99%,...) because – normally – scientists rely on statistical methods, and they use samples to represent the world. They also try to isolate a few factors and study them, and this leads to an important caveat.

Successful science should – in the light of a wider and systems-inspired perspective on reality – try to avoid the "single-cause" fallacy.[25] It should be cautious of using a version of reality that is so simplified that it approximates the world too much, leaving different causes and variables out of the equation.

The approximation associated with hard (natural) sciences is seeing the world as a limited paradigm, and missing different causes and bigger pictures, leading to slower evolution and narrow perspectives. This danger is a feature of the system as it helps focus the energies on nearby difficult problems, but it carries within itself the seeds of inefficiency and illusion.

If the hard sciences tend to focus on small problems, and thus have limited perspectives and slow incremental progress (reduction approximation), we can see the flip side in the social sciences (generalization approximation).

Complete social science fields are sometimes built on (obscured) sweeping generalizations. Approximations roam free and learners sometimes miss the real factors causing what they're studying, as the real cause might dwell on another "macro" plane. Entire fields of inquiry like economics or strategy are based on a set of approximations and generalizations that can lead to attribution errors or postulates unusable in the complex environment which these sciences were originally intended to explain. Sometimes, it might be useful to just "know more," but the practical value of a science can be lost without an anchor.

The field of strategy for example presents many great models, but to me, it always felt most useful when it was about specific details about specific situations (anecdotal, soft, developmental), instead of stuffing heterogeneous stats into a Frankenstein-like model.[26] It can be more instructive as a catalyst of flexibility and action, and a reflective space. Otherwise, it is just too generic and anti-action (ironically).

The field of economics offers an interesting perspective on modeling, but generalizations assuming rationality and abstracted interaction models can be both wasteful and harmful. The assumption that these are proper approximations of some real limit (or worse, situation) has no proper realistic backing. The difference between isolated interactions of rational agents and real situations is not quantitative. In summary the value of this line of inquiry might not be economically justifiable considering the effort needed to create real utility for it. The scale of the unseen approximation is too great, and the utility of its outcomes is highly questionable.

Why are these fields so popular then?

Here again, the social and the cultural – not the scientific – are the driving aspects. Science is a signal sometimes.

Some of these fields maintain a disproportionate level of attractiveness and prestige (compared to their returns) because of their target audience. The market – as we discussed in the previous chapter – works in mysterious ways. These sweeping-generalization fields of study are useful as legitimacy tools for powerful agents (the target audience), and for justifying already-taken decisions. This target audience of "powerful agents" might include CEOs, political leaders, administrators, and other "high-level" leaders who need these kinds of "aggregation sciences" to feel better about doing their jobs, make others feel better about that, or for appearances.

This function of some sciences as signs, or as "rationality and legitimacy" jewelry should not be shocking. The university itself, as an institution, has an awkward relationship with that function.

I am not making any value claims here. Displaying signals of legitimacy and rationality are important leadership tools and necessary practices. My objective is to point out the different layers of approximations included in these fields of study, and their entanglement with social, cultural, and political variables. Practical results are a separate (and very substantial) discussion.

Between excessive reductionism, and sweeping generalizations, the sweet middle seems to be mindlessly lost in a mysterious fog.

Acknowledging the ever-present scientific uncertainty, and maintaining an awareness of the different reductions and approximations that the scientific process, and system, include, are equivalent to the conscious awareness (mindfulness) aspect in the overall flow diagram presented in the early chapters. This is what determines whether the quantum outcome is likely to be on the dark or the bright side.

Without keeping this awareness in mind, science is prone to fall victim to the darker blade, and its "devotees" risk being misled into the prison of illusion and conflation. Additionally, the trap of becoming loyal to your camp (for many reasons), and not to the process or the search for truth, awaits the unsuspecting. It is easier and more satisfying.

Science as ego and politics[27]: faces of the social process

I mentioned before that there were different "camps" in science. If things were about certain truths, there would be no need for opposing camps (that is, if we assume that people are rational and good).

How do the people in the different camps choose their side?

Is it always careful examination, devoid of any subjective inclinations, peer pressures, and cultural factors? In other words, are all of them as rational as economists? Is it not dependent on "initial conditions" (some form of the anchoring bias) where the point of origin (where you started) has a greater-than-expected effect on your ultimate destination?

We can't answer the above with certainty.

There are more social and political elements within science than meets the eye.

Paradigm shifts are – at least partly – a social process. What is included within these social contexts?

Imagine you are a well-respected scientist who has spent a career working within a certain field, teaching students, authoring research, and supervising it. You've followed a certain direction or set of principles, and adhered to a set of accepted methods.

Do you think it will be easy for you – subjectively – to accept proposals that detract from your chosen direction and legacy, and maybe discredit part of your work?

Now let's take the other perspective. Imagine you are a young and new scientist. You are trying to get a post within a university, and the university will only give fixed posts to a select few. The selection criteria often include something like publishing several research peer-reviewed articles in academic journals.

Do you know who the "peer" is in the previous sentence? Well, it is the well-respected scientist from the previous paragraph.

What should you do now?

If you try to work on something that truly inspires you and motivates you, and that is creative and satisfying, you're in trouble. The well-respected scientist will probably reject your work, because that is his task in the system, and because academia is conservative, and because...momentum.

What do young scientists actually do?

Usually it is "let me publish what is more likely to be published." "It is a 'publish or perish' world out there" (a dreadful academic cliché). I remember a professor telling us that it would be wise for academics to leave big and revolutionary projects for later in their academic careers (?!?!?!), and at the beginning to focus on "things that get published" (I'm still not sure which annoyed me more, that sentence or the staccato finger-snaps that accompanied it).

A new scientific truth does not triumph by convincing its opponents and making them see the light, but rather because its opponents eventually die, and a new generation grows up that is familiar with it.

Max Planck [28]

Harsh but true.

"Later" rarely – if ever – comes.[29]

Science is uncertain most (all) of the time. It is a social process, embedded within a social context. Scientists are people who might be subject to biases and prejudices, even in the context of their work (science), and they are working people who have social and economic burdens to worry about.

Our vision of science as an ultimate truth is itself a convenient approximation.

Most of these qualifications go unseen most of the time, and we approximate science as a utopian endeavor, and scientists as saviors (or the opposite).

We should also pay attention to the fact that science, as an organized and institutionalized effort, can be considered to be a representative of different power structures within a society. Science will represent social and religious groups (because scientists are human), big businesses (because they donate to research and universities), and other political affiliations (for any reason).

Increasingly, people are using science as a component of their "cultural identity" and positions on scientific issues sometimes are caused by ideological inclinations. Examples of this include vaccination and autism, evolution, wearing masks during the Covid-19 pandemic, among others. The positions taken by different groups of people are frequently more expressive of political statements and declarations of political affiliation than they are scientific.

Let's say I'm a "progressive," and I want to express this affiliation. I will take an almost ideological position in support of what some scientist said on something, and fanatically (religiously?) defend it, belittling people who question it. Or if I'm conservative, I might be too aggressively skeptical of some scientific truth, refusing to understand how some statements fit within our overall epistemic system.

In both these cases, people's positions are not scientific. It is the "ego" or the personal identity (imagined and constructed) that is being projected, praised, and defended by these (apparently) scientific positions. People might see attacks on a certain statement as an attack on their cultural identity.

True understanding and evaluation of the issues is approximated as (replaced by) a set of biases that stem from the need to defend the self and its social and cultural affiliation. Attacks on the self, culture, and values can be quite distressing, so conflicts arise and a vicious loop of struggle gets nurtured.

The real problem of the ivory tower might not be that it is an ivory tower. It might be that there was never really such a thing.

The problem(s) with experts/scientists

One more approximation that goes unnoticed here is that the scientist status is used as an approximation for being an "expert" in a field. This might – or might not – be true, but it definitely raises many issues.

Statistically speaking, scientists are smart. It takes intelligence and academic distinction to reach advanced stages of education[30].

The fallacy (Halo Effect), however, is in assuming that this intelligence grants super-human abilities, and that their "procedural intelligence" can be generalized to different skills and competencies.

The research on this problem isn't lacking, and there isn't a shortage of research and statistics that show the limitations of intelligence as a predictor of success or real-world achievements, which tend to rely deeply on a wide set of emotional and social skills. I can imagine this problem playing out in the different sciences, but especially in the highly technical ones, where more and more specialization into micro-scale phenomena is required to make meaningful contributions.

As we saw, scientists are subject to group prejudices, and – normally – they have emotional and social considerations to tend to. They tend to work within a paradigm, and – being human – could be prone to mistaking the paradigm they're stuck in as reality (nobody wants to admit that they live in a narrow and approximated portion of reality, because that is less cool than living in the "one true" reality). They are very familiar with that particular paradigm/bubble, and with the people who have lived in that bubble, dead and alive ("We like what we know," "You become what you understand," etc...). It always astounds me how much academic and research effort goes into the details of the tradition and the minutiae of the other researchers who've been through this particular path: these are indirect tools of indoctrination.

The excessive linearity of the hypothetico-deductive method can become a serious problem. It can hinder a cross-disciplinary understanding and an opening up to the methods and "knowledge" in other fields. It can encourage the fallacy of approximating the world as the set of phenomena directly involved with a specific area of work. It can distort the three-dimensional (or n-dimensional) world into a linear projection of itself.

Expert scientists are not categorically superior to the population on account of their academic position alone. We only think that because of a series of approximations, within science itself, about the scientists' role, and about the social contexts around science and scientists.

The technocrats will fix things (they won't)

We attribute competence, integrity, and craftiness to scientists/experts, and we assume they know what they're doing. Here, it is interesting to go a little bit into the "technocrat" as a good leader image, where people in many countries (including mine lately) seem to think that their problems would be solved if "technocrats" somehow occupied the political leadership positions in the country.

Nassim Nicholas Taleb in *Antifragile* (2012) talks about the IYI (Intellectual-Yet-Idiot), and I'd like – in the spirit of more gentle, and less approximate, name-calling – to talk about the "NADE" (Narrow and Depressing Expert).

That's someone who – even though intelligent – has been put down by academia and forced to research things that he/she doesn't really like to get a publishing record, thinks they are too smart compared to others in society, and so tends to be negative and bitter about it. Their interests, due to competitiveness, chance, or mistakes, haven't been expanded beyond a limited area, and so they have wasted the true potential of their mental resources. They can't believe that with all their skills and intelligence they get relatively small material rewards. The limited perspective contributes to a particular set of traits, too, that translate to "seeing everything as a nail."

Technocrats, many of whom are closet NADEs, can't – generally – solve big complex problems without upgrading their operational toolkit.

A minister responsible for power and energy in a country within a financial crisis, with frequent power outages and shortages, is a great example. This is an Expert/PhD in the field of electrical engineering who has – under uninformed and unfortunate public calls for "Technocrats" – been given the responsibility of running the ministry.

In press conferences to explain the situation, the guy talks about generator capacities, specs of storage facilities, and terms and conditions within a tender to purchase technical equipment. With a numb and unsympathetic face, recounting numbers in units most people don't understand is the only answer he could find to solving a giant problem which has different social, economic, and political manifestations.

It is natural that his sincere efforts were not appreciated.

His technical fixation on the small-scale nuts-and-bolts is a continuation of his work in research and consulting, but that is hardly a qualification for leading.

NADEs can be thought of as being somehow similar to IYIs, but I'm trying to present them in a different way. NADEs are not necessarily:

- Intellectuals: I think the status of being an "intellectual" goes beyond being a scientist/expert in a field. An intellectual is not a PhD holder. The intellectual is an explorer of knowledge and a lover of wisdom, and different social and natural sciences. They are a seeker of truth.

- Idiots: As I said, many are really smart (on paper or on a computer screen).

But approximating them as intelligent in the sense of managerial/leadership competence and general problem solving can be a drastic mistake.

They are narrow, extremely linear, and very depressing.

* Bilbol's Curse

(Or: Say the truth)

"I noticed that you're always doing this. Changing everyone. Removing some of your existing workers, and adding new ones every few days. Why do you do that?" Grumis asked once.

"Yes...I have to. They can't work here for long. After all this time, I'm the one who's been here the longest. Everyone else leaves," Zif explained.

"What prevents them from staying?"

"It is an old tale. I'm not sure if the explanation is reasonable, but there is probably a hint of truth in it. It is called "The Curse of Bilbol"...many problems really. Didn't you notice anything strange when workers spend a long time here?"

"Only that they change periodically"

"That's probably because you didn't try to hear them."

"I do hear them, but I don't understand what they say...I don't understand their language."

"You mean 'languages'. They don't speak one language."

"Interesting."

"Yes...They start off speaking the same language, but as they spend more time on the tower they change. Some of them say that this is the curse of Bilbol, who had built the original high tower of the mountain. He didn't want anyone to reach the height of his achievements ever again, and so he condemned all who try to build a higher tower to lose their touch with reality as they spend more time on that project. As time passes, their language changes."

"... and they start speaking differently?"

"Yes. They start speaking differently, probably because they're thinking differently. They start saying strange things...The more superstitious ones actually believe that Bilbol is the soul of languages, and he is the one speaking through them."

"Fascinating!"

"As they spend more time on the tower, Bilbol has more time to control them and start speaking through them, so they effectively lose touch with their colleagues. This is why it is better to take a break and leave the place before the effect takes hold of you. It is strange because it usually starts quite slowly. Just a few strange words in the middle of a sentence, and which have no clear meaning. It starts with a few new words that they add into proper sentences."

"and then...?"

"The number of strange words starts increasing. What starts out as isolated and limited, later starts to multiply. In some of the advanced cases in the beginning we could still hear a few proper words, but it seems that they became disconnected as the new language took over. Later on we started intervening earlier, before it was too late."

"Couldn't you understand what they said? The ones who start saying different things?"

"Well it wasn't actually coherent. We listened and wrote...they don't say similar things. It is not one different and new language. More like gibberish to us...Probably each one expresses a different unique language – no way to tell. We were very limited on what we could actually do to solve these problems. Was it really meaningless, or was there something – in them or in the place – trying to talk to us...? That we will never know."

"So there was no way for them to continue working as they lost their original language?"

"Yes...Would you be able to? Imagine the slow removal of everything that holds your sanity and personal history as one piece. Sometimes they'd start becoming unpredictable and very emotional. Some of them burst into tears if they heard something – we couldn't learn why – and some just became furious unpredictably...Some broke into laughter. It was something we couldn't manage. Letting them go was the only option. If they don't go, their words will. If they don't keep their words, they don't keep their minds."

"True."

"If Bilbol really steals their language and replaces it with his words...I would have loved to know what he wanted. Sadly, people without a language might be saying smart things, but they can't build well with others."

The language prison: implicit assumptions and the group's DNA

We ourselves have been created by the invention of a specifically human language. As Darwin says (The Descent of Man, part 1, chapter III), the use and development of the human language "reacted on the mind itself". The statements of our language can describe a state of affairs, they can be objectively true or false. So the search for objective truth can begin – the acquisition of human knowledge.

Inazo Nitobe[31]

Language as model: networks of (approximate) signs

To remain loyal to the language we're using in this book, we can say that language is a system (some would say structured system) of symbols (signs) that is used for communication. It is one of our cultural subsystems.

When thinking about language, many people might consider it a tool for self-expression. The more accurate reality is different. Languages do serve many purposes – true, but – and considering their origins – languages can be better thought of as tools to facilitate cooperation, rather than means to express ideas. [32]

Language is another great model of the concepts we're trying to illustrate.

A language's words are most often embedded within networks of concepts and other words. The links between nodes in these networks can be related to objective meanings, personal experiences, culture, syntax, etc...These networks – overall – help generate the meaning of a certain word, as well as its "connotations" or "excess meanings" (so, what else does it mean or relate to, beyond the direct equivalence?).

Words, and the subtle connotations and hidden meanings enveloped in them, can be thought of as approximate placeholders for concepts in speakers' minds. Words fit concepts and convey them easily most of the time (a significant part of the human experience is shared), but sometimes the fit is not exact. These words have, for example, unique characteristics[33] that can't move easily across languages...Words are not just descriptions of the concepts, but rather add different historical, personal, and cultural "colors" to them. So just as the same image can be colored in different ways, or passed through different color filters, the same word can have different colors that are perceived by different people or in different groups/cultures.

This helps explain why translations sometimes feel awkward and wrong, and many digital advertising banners appear so cringeworthy (specifically interesting for languages other than English).

The implicit relationships get lost in translation.

Overall, words are approximations for concepts, and they serve that function by relating a certain concept to an existing network of concepts and ideas (other words and phrases). Representations linked to stories, peculiarities, and subjective experiences acquire specific emotional and cultural nuances (of personal and group origins) with time.

Language(s) probably evolved to solve simple coordination problems. This original function shaped (still does) structures and syntax rules, as well as other aspects too. But as the scope of language expands to more sophisticated artistic and scientific expression needs, coordination problems are bound to multiply. This is especially true because paying conscious attention to all the excess meanings contained in words is a seriously demanding cognitive task.

Incompleteness and complexity: words of doubt and defense

English is a surprisingly nonspecific language, and the multiple meanings of common words often trips up our ability to understand what's being discussed.

Seth Godin (2020)

The problem of perpetual incompleteness, discussed in the "Inevitability of Incompleteness" chapter, has a special relationship with the nature and uses of language.

Even though languages are meant to be practical, many communication goals can never be achieved properly with the linear-logical system of common symbols that is language. This can be attributed to people's subjective experiences of language and reality, and to the fact that the nuances of a certain language can "color reality."[34]

Words have a history, and that is contained in them. They also serve special roles within our internalized cultural resource sets. They sometimes fail (or come short) in expressing different subjective concepts, or complex real ones (failure is relative). All that, added to the inaccurate use of language (or the improper use of inaccurate language) can drastically increase complexity in social situations.

This is a prototypical vicious loop: if there is complexity in a situation, the use of "loose" language can add a layer of complexity because the original complexity requires more explanation, and complexity will keep spiraling up (think of the "add complexity to complexity" effect that many will fall into as they try desperately to replace one failed explanation with another, or as your argumentative friend feverishly jumps from an illustrative example to the next to try and clarify one idea)… after a few rounds of verbal (or non-verbal) exchange, we can reach a phase of emotionally charged self and group defense, or a frustration-ridden inability to communicate!

Some examples of situations where languages are likely to increase misunderstanding because they lack the necessary tools can be:

In Art/Creativity:

Someone trying to express certain creative/artistic visions can get their work interpreted by authorities in their community as "dangerous" or "blasphemous" or challenging to high powers (godly and godly-representative). Novelty here can pose a challenge to the network of implied meanings and collective assumptions, even indirectly. Spiritual people were (are) sometimes persecuted (as heretics) by religious fanatics and religious authorities because of these (non-objective) verbal chains of influence.

In Science/Thought

Someone trying to frame a new scientific understanding of a certain phenomenon, but because of the yet-inexistent network of concepts that need to be connected to their new language, can get their ideas rejected (remember the context for scientific revolutions) even without proper consideration. The terminology and supporting network of concepts isn't ripe yet.

In Social/Political Contexts

The discussion of certain social and political (and even scientific) issues gets derailed into conflicts because of associations. Some concepts are linked (naturally or purposefully) to people's values, even though they could be approached from a purely practical and reasonable perspective. Many will feel their stability, identity, or group is being threatened. Discussing healthcare for the population, monetary freedoms, vaccination, and even the role of science, gets morphed into activities of feverish ego-defense and group-preservation.

The identity problem (again): who's saying whom?

Are you talking when you talk, or is your language (and culture) talking through you (or, more dramatically, talking you)?

Words' "subtle" meanings and connotations are (mostly) culturally-inspired. They carry historical and social significance, and they sometimes chronicle trends and efforts by groups to preserve and promote certain meanings (values).

Language is not only a tool to transmit our ideas or to reduce the costs of cooperation. The specific patterns in a language, its recurring structures, and embodied history materialize as words: language frames and affects thinking (individual-internal) and collective action (group-external).

Certain statements, patterns, aesthetic preferences, and cultural themes can be more likely in certain languages and cultural environments. This extends to personal goals and values: what individuals value and seek in life is affected by what they understand, which is itself shaped by the tools of that understanding: words in a certain language.

In a sense, the language shapes people's thoughts and actions in specific directions.

The discussion of language-as-an-identity-builder relates strongly to our discussion of the DMN (default-mode-network) and passive thinking. Most people, in fact, "hear" their internal monologue in spoken words, in their own voice.

Our passive ongoing thoughts and internal monologue – usually – use the same language we use in our everyday speech. As they do, they are continually laying the foundations for our habits and wants in the specific patterns of a certain language. These daily habits, thoughts, and aspirations are our future self![35]

The charming glue of bosons: nouns of destiny

I was watching a lecture on the citizens of the microcosm: the (very) tiny particles that are the building stuff of tiny particles. I find the topic fascinating, especially considering its philosophical implications.

The problem with this particular video, however, was with the way the presenter (an esteemed scientist for sure) was delivering his explanation. The bulk of transmitted knowledge in this video was just words and their meanings. The terminology and dictionary parts were the center of the discussion.

Words, terminology, and knowledge are closely intertwined. I concede that knowledge – largely – lies in the acquisition of vocabulary (words) within a certain field (language). These words – as cognitive tools – through their implicit relationships and unique properties will facilitate knowledge exchanges and the quest into deeper knowledge within that field. They frame thinking, and can also serve the (bonus) function of scaring away intruders (our field, our words, our language).

Here, two areas of concern appeared.

First, are we always aware of the extent to which the "excess meaning" or "unattended cargo" of the words can affect the trajectory of evolution of the language? Stated more clearly: can these "original terms" lock us into a paradigm that constricts (rather than facilitates) our knowledge quest? In this sense, our initial weaknesses and myopia will be the seeds of enslaving us longer into a distorted version of reality that works correctly from a limited perspective only. This relates to Kuhn's propositions, but on a more radical and subtle level...

Second, shouldn't a knowledge quest be more focused on fundamentals and relationships rather than terms? Essentially here (and not in general) I am somehow arguing in favor of adjectives and verbs as opposed to nouns: "Waves" (and processes and states) as opposed to "things." Why do I care about things being called "Charm," "Boson," and "Gluon"? The names themselves can be sources of entanglement into a web of incompleteness and limitation...What if we later need to replace all this perspective?

Nouns here are doctrines and belief systems! They are too confident in their preciseness.

Let's build "better" languages (should we really?)

Some people prefer the language of mathematics to solve problems and deficiencies that come with trying to use traditional languages in science. Mathematics is a precise language and an exceptional "formalization" tool: it is the "objectifier" of reality. But it is not without weaknesses. Precision is frequently associated with incrementalism, and moving slower. A mathematical system also will inevitably be incomplete (Gödel), which can be quite misleading.

Still, programming languages are an example of formal languages with (significantly) reduced ambiguity. Logical statements are translated into an algorithm (recipe or instructions) or pseudo-code. These are then transformed into code that specialized software packages will translate into simple instructions that can be understood by (machine) processors. There is an elegance and beauty in the precise expressions of algorithms and pseudo-codes. Writing for machines can force increased clarity and the exploration of different thought and description patterns (like recursion and other logical properties)...

Many have tried to create new languages altogether... Still, Wittgenstein commented that a language that developed inorganically is not only useless, but despicable.

Engineered or constructed languages are languages created with a specific purpose, like serving some philosophical or logical function. Of those, there is a group of languages (logical languages) that have been created precisely for the purpose of enforcing unambiguous statements[36] – this is done by enforcing certain expression structures and sentence syntax.

Note: The most popular constructed language (Esperanto) was intended to be a "universal second language." It is based on European languages, but is grammatically light and scalable (easy use of prefixes and suffixes).

Lojban (the successor to the first attempt: Loglan) is another example of a constructed language, and significant work has been done with it. Its community believes that it can be learned more systematically, contribute to AI work, advance scientific knowledge, and even open horizons for cultural creativity.

This is an interesting quote about Lojban's perceived potential:

We haven't, though, tried to impose a system on the universe like most a priori languages have. Instead, we have tried to broaden gismu flexibility so that multiple approaches to classifying the universe are possible. Our rule is that any word have one meaning, not that any meaning have one word. There is no "proper" classification scheme in Lojban... Lojban offers a new world of thought.[37]

I'm – personally – not sure if this will ever work, or if it is a great idea to push the "creative ambiguity" out of language (although it might give rise to new forms of creativity). All this is highly dependent on the context and use-case, but contemplating the ways in which languages can shape action outcomes points to the issues that can be raised by the approximations inherent in language.

Religion discussions: the original sin

The follies of approximate language use are also visible in the theist-atheist debate.

Are these people trying to express completely logical positions? Can you imagine someone defending their god and remaining rational in their expressions? Can you imagine a language so value-free that it can carry this discussion?

All I could see and hear when I spent (questionably-regrettable) time listening to different debates between theists and atheists was a mess of concepts (In my defense, those were my hobbies when I was much younger, more handsome and less wise – if that is possible). The jousters keep on jumping around between certain words, regurgitating worn-out ideas, and marching in tragicomic speech sequences.

But why shouldn't they?

One is trying to murder god, the other is trying to save god. Both tasks are quite emotionally involving.

Without going into too much detail, most of the popular attacks from one side are against some mental image built on an "old man in the clouds" metaphor, or arguments orbiting infantile definitions of fairness and determinism.

Not to be outdone, the other side argues sometimes against plain scientific truths or (if they're smarter) against some purposefully distorted version of these truths to keep a certain naive story passable as long as possible.

The language used is not suitable (remember: I am talking about the public discourse around this). It makes it easy to get lost in generalizations or to add an emotional component, or a very significant self-defense and value-defense component to the mix. This is because ideas from both sides are approximations for something else, which itself is an approximation, and so on.

Belonging, the self, ancestors, and values are getting crammed into the discussion.

I've read *The God Delusion* by Dawkins (2008) (again when I was younger and less wise), and it felt like a "NADE" product. Still, extremely ethno-centric (tribal / limited / racist) religious thinking is very dangerous, and one can empathize with the sensitivity from the other side.

It is also dangerous (in a different sense) for a discussion about "God" to use as the prototype of "God" a simplistic anthropomorphic image which wrestles with other men, harbors uncontrollable rage, and encourages the casual slaughter of "others." This image is what comes to the minds of many when they think of god (connotations), and so they are bound to have a negative reaction to it. They, after all, usually don't appreciate the fact that this particular image works on a particular layer (or slice) of reality, and that layer is different than the one they're looking from. After the negative reaction, the science can be conveniently adjusted to support one vague statement versus the other.

This conceptual soup happens because the discussion isn't a real discussion about god and spirituality, but name-calling and self-justification escalated by the mess of meanings contained in approximating words!

"Muslims are Terrorists"

What does it mean to discuss the link between a certain Islam (mine let's say) and terrorism? Let's examine it under the language-approximation lens.

A group of people, who belong to a certain group (technically a subset of Islam, not that they agree to this), have committed acts of terrorism and murdered civilians.

There are three layers of approximation possible here:

1. Is it reasonable to approximate these (murderous) people as representative of their own group?

2. Is it reasonable – if you agree with 1 – to approximate their own group as representative of Islam in general?
3. If so, is it reasonable to approximate other Muslims as belonging in the same set?

What if different versions of Islam differ drastically? How is national identity integrated into this mix? What political factors come into play?

Is this terrorist more a Muslim, or a (insert nationality), or a (insert socio-economic variable), or a (insert religious sect), or a (insert psychological profile), or a (any other belonging which can be measured to correlate with "terrorist" traits)?

Why is this word "Muslim" considered fully and supremely representative of someone's identity and why can it render all other competing dimensions in it mute (especially if most of this guy's victims are Muslims)? Does the problem lie in the difficulty of finding a more "discriminating" word?

No need...this approximation is "good" enough.

Approximations are convenient...And they are everywhere!

Approximations make it easy to create the convoluted political mess that facilitates the use of religious identities in all kinds of political (and social and economic and personal) projects!

Institutional inefficiencies: the systematic organization and leadership of groups

Wicked problems: systems of (systems of (systems of (systems of ...)))

In our effort to examine the impacts of absent-minded approximations, the domain of social systems and institutions fits (roughly) within the Collective-External quadrant of our AQAL categorization.

This domain is particularly complex, vast, and hard to navigate.

The term "Wicked Problem" was used within urban/social planning (later moving to the design field) to define a specific set of (especially problematic) problems. Wicked problems are hard to formulate and define (can be phrased differently by different people), they don't have one correct solution, and they interact with other problems[38]. They are really complex, and their complexity is usually a symptom of the interaction of different systems within a super-system, example: every individual is their own system of emotions and thoughts, then groups are systems of individuals, then communities are systems of groups, etc...

Many different polarizing political, social, and economic problems faced by societies are wicked problems. Think about the problems of social injustice, environmental sustainability, and cultural policies. Should we increase taxes? Should we prioritize economic growth? Should we aim for income equality in our economic/political planning?

These problems can't be easily reduced (except if you're one of the "it is obvious to everyone that..." people) or solved. As you attempt to solve them, you will be surprised with the rise of other unexpected issues.

Sarkar and Kotler (2020) have outlined an interesting "ecosystem of wicked problems" to illustrate how a bunch of these wicked problems relate to each other, and that can give an idea about the level of complexity involved here.

Wicked problems are not just linked to the political and planning fields. They are rampant in business, management, design, and work in general too.

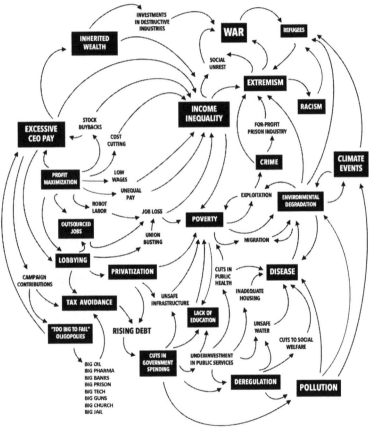

Figure 13: The Ecosystem of Wicked Problems
(Sarkar & Kotler, 2019)

Tragic approximations of mega (wicked) problems – Why?

So how do we deal with complexity of this scale? (Spoiler: you know how).

We replace some issues and approximate others till somehow (to someone) satisfactory progress is made. This is what "satisficing" means: we reach a temporary state of equilibrium which is somehow acceptable – it is acceptable within the bounds of rationality, resources, and contexts.

But why is this tragic?

A corollary of the principle of inevitable (perpetual) incompleteness is that we are always in an unfinished and evolving (work in progress) state. This means that the best we can do is have our solutions to these problems at a particular time be a reflection of our capabilities and needs at that stage, right?

It is not what we do. Inertia and momentum are stronger than you think.

How do we solve the transportation problem? Is it reasonable to use one to two tons of metal and plastic to move – on average – 1.5 people around?

How do we solve the power-grid problem? Is wrapping the globe in wires the most effective way to have power?

It might be what's convenient, but it is not necessarily what's best.

We get stuck with certain solutions way beyond their ideal shelf-life. Our current solutions are what we have carried forward from the past (give or take a few tweaks). They are past approximations with makeup.

The surviving past is great, and old wisdom has survived because it is "wise." It still has to be qualified.

It is convenient not to change things as any self-respecting bureaucrat would tell you. In fact, "not doing anything" about a problem is often a surprisingly powerful strategy.

But the linearity and slow rate of evolution of our solutions create inefficiencies and waste. These increase with time. The equilibrium that we think exists is an intermission as greater chaos silently builds up.

The problem is that the lingering of solutions for too long (like a political system that hasn't been updated in 40 years) might make our "current" reactions to wicked problems more reflective of the past and of "other things" than a more ideal reaction, based on our current resources.

What are these "other things"?

They are the items and concepts that implicitly sneak into the problem with the approximations we've originally used to solve the problems. Our approximations are a reflection of prejudices and preferences (in this case they are collective/group preferences, which might reflect cultural differentiation and idiosyncratic histories or specific "initial conditions").

Some political and social institutions might be – obviously – dysfunctional and inefficient, yet can't be replaced because such an act requires a certain tilt in the power balance in society, and the cost of keeping them, even though high, might be perceived as lower than the cost of the "worst-case scenario" that might happen if we attempt to change them.

We settle for the bad, because the unknown might be worse. (So variance becomes more important than the expected value here).

A famous line in an Egyptian play goes: "the boss we know is better than the boss we don't know." (It is funny in an Arabic-Egyptian cultural/meme context.) The inevitability of this is the tragic part. We stick with the boss we know for way too long!

Political representation: approximating the "Will of the People"

Why is populism a great concern now?

Why is "identity politics" a great force?

Why is "fake news" a real thing?

The existence and functioning of political institutions, which are supposed to lead human societies, is subject to a number of approximations and assumptions.

Political problems are notoriously wicked. They are very complex, their participants are in a state of incomplete knowledge, and they remain highly contested and charged as they reflect a struggle of personal and group interests, culture, and identities.

Political authorities are assumed to be an accurate-enough representative of the will of the people, or of some "higher" or "more noble" power (like God or big business or X's family or clan). Let's focus on the first (the will of the people) since the second form often has the assumption that this "higher" power is a more refined version of the will of the people.

Check approval ratings of presidents and parliaments after one or so years of elections[39] to see the problem. These institutions aren't really approximating the will of the people, but rather "filtering" it somehow (for better or for worse). Is it optimal to be governed by a representative of 20% of the people? It is better than the alternative (?).

Democratic systems are supposed to reflect – with time – the evolving will of the people. The problem is that the lag of this process is not negligible (so the process is slow and evolving, with feedback and complexity), especially when compared with the speed of modern communications and the volatility of human moods.

The "will of the people" is a very vague concept that gets transformed in weird ways into power structures in the state's management, utilizing all kinds of identity-related illusions in the process. (Link this to the "I am (We are) great" discussion.)

In a sense, politics can be (rationally) thought of as an exercise in priorities. Many problems would be solved if

"Priority Politics" was acknowledged and treated as such. At a deep level, we all (except for cartoon-style villains) care about similar things, but the order of importance of these big issues is what differentiates us. Mixing up different causes and identities (or identity-illusions) with positions on many economic and policy issues is lucrative business, and a great shortcut for (professional) politicians. It is a clear (to some) exploitation of approximation.

Causes conflated with coins: on one level, you'd have people fighting (and killing, and getting killed) for their god/culture/ values with zeal and rage, and on another level, you have people bargaining and negotiating technicalities and spoils with (apparent) civility.

Another inherent approximation here is that people mistake (approximate) career politicians for leaders. People think that leaders should be principled, sincere, caring, intelligent, and wise[40]. Think of your politicians: how many embody these traits?

That's not strange – what do career politicians (most of the time) do? Usually their job is to manage daily requests and favors for a large set of people and engage in laborious and non-ending give-and-take negotiations. This shifts their perspective from the long-term oriented traits explained above into another set.

This itself is another approximation: What horizon do we have for results that we're trying to achieve? Are we looking for long-term results that affect the structure and identity of our people, nation, and groups, or for shorter-term ones that have to do with specific power gains and economic adjustments? The two options can be harder to relate than most assume, and the question goes unanswered. Everyone assumes their own answer.

Can technology solve (some of) this?

Because inefficiencies of communication and understanding (information distribution) lie at the heart of many of these problems, we have to ask: can this be solved by technology? Part of the way political systems are designed has been carried forward from the past (inertia): due to the difficulties of checking people's opinions on a variety of issues, and because they might not know what to think on something, we approximate them.

We approximate every 1000 or 1,000,000 people with a representative (they elect him/her), who then takes decisions for them (representation) or gets in touch with them occasionally. What if – through existing information-communication-technology tools – we can get people's opinions directly on a specified set of issues (even if limited) and drastically change the function of political representation. This isn't as hard as it used to be when the current institutions were conceptualized.

Many issues can be decoupled from "identity politics," reducing the room that politicians have for utilizing approximations to conflate issues of identity, patriotism, and other higher values with practical decisions like environmental policies, urban planning, science, inflation-management, and investment decisions, and on...

People can instantaneously and directly vote on something (preferably a yes-no) and "cut the middle man (woman)." This idea is promising, especially if refined by considering knowledge, qualifications, and the disproportionate risk/relevance of decisions to specific groups. This doesn't mean that party-politics should go extinct, but rather points to the fact that a more direct representation has been more feasible, but not sufficiently explored because of the past's momentum (and the interests of the decision makers). Set-ups that might free political leaders to do more "leadership" and less management need to be considered.

"How lazy is John Galt?"

Ayn Rand's books spark many discussions, and they are distributed (for free, and otherwise) widely. Some surveys place her most important book, *Atlas Shrugged*, as "the second most influential" book (after the Bible).[41]

Disclaimer: I didn't read the book *Atlas Shrugged*, but rather watched the three films [42]. I am a hungry reader in general, but I'm stingy with time. *Atlas Shrugged* (~1000 pages) seemed too heavy a load, especially after reading *Anthem*.

The big idea of "Atlas Shrugged" is that the really creative, rational, and hard-working *individuals* (men of the mind) in society can stop the wheels of civilization if they stopped working. Everything, after all, depends on their efforts, dedication, and genius, and the others are either too lazy or too inconsequential. Society drains the energy of its most productive members and abuses their sacrifices, so what will happen if they just disappear (into a mountain) and go on strike?

There might be some (faint) logic in the premise, but the story and the events are very problematic. The major theme – I would say – is not "objectivism" (Ayn Rand's philosophy, which she uses this book to introduce), but an unbearable *conflationism* of many different concepts and ideas. It isn't really clear (to the author, or to us, the audience) if the wondrous "John Galts" of society are actually inventors or innovators, hard workers or good managers, scientists or leaders. This is because the lines between these different functions have been blurred by (for) the writer. A (somehow rudimentary) understanding of how organizations are managed, how innovations happen, and how systems of work and production interact could have saved the world a lot of confusion (and twitter many threads).

True leaders don't just "invent a metal," but work with different groups of people to achieve progress. It is unlikely that if they run away to some mountain, they can create a utopia with other like-minded Galts...Productivity and innovations happen within systems. Geniuses are very important to progress, but to assume that these geniuses of Ayn Rand (business leaders, mostly) are actually the geniuses she thinks they are (inventors, leaders, and movers of civilization), or that they can function on their own and create their own utopia is contrary even to the likely reason for their success (working and cooperating with others).

Some of the logical mistakes of the story are really infantile, and I invite the reader to sample the movies. I admit that I was a bit too annoyed when they went into the magic place inhabited fully by talented people, to find out that that place had dishes, houses, glasses, and cars (so my judgment was clouded, and I could be approximating my anger with this "seemingly rational" critique).

The discussion of the book's style or its representative validity isn't my point. I included this example as a (popular) comment on the usual misattribution of outcomes to specific causes (e.g. person "X" succeeded because he is a genius hard-worker), and the reduction of systems to individual agents (e.g. person "Y" built a rail-road): both are symptoms of dark approximations. Moreover, the (apparent) market success of the book (and others) in terms of distribution and readership is vastly disproportionate to its entertainment or intellectual quality.[43] This is because the book is more of a political statement than a book, and in that sense, the conflation and reductionism make perfect sense.

The organizational soup: leaders, managers, and others

A similar version of the political representation problem is sometimes encountered when discussing the management of publicly traded companies. The pressure on business leaders by investors and representative boards with short-term profit targets is to drive quarterly profitability and growth figures. This sometimes creates a conflict with sound long-term planning, which might require weaker results now. Many solutions to this problem have been proposed, but many business leaders prefer to solve the problem by avoiding it altogether: stop being a public company.

Company executives should have (ideally) the optimal blend of leadership (vision, change, transition, progress, long-term) and management (execution, stability, efficiency, short-term results). We think of them as both, and so do their shareholders. This – in reality – is just a shifting set of approximations (the conflation of leaders and managers is a very convenient approximation, and good business for business schools and consultants, and a good assurance for political leaders too).

Can we further break down the leadership and management functions into sub-roles and distribute those? Maybe, but that will be the task of a really courageous organization, otherwise, things are okay as they are...We have the "least-bad" solution.

How do managers manage? How do they know what decisions to make?

Managerial Approximation Styles

Gut-feelings are a big part of it, so is "scientific" management and grounded strategic initiatives!

Organizational leaders get an understanding of the complex environment by mapping different factors into their (preferred) mental models (their mental models are how things are represented in their minds). This means they understand the environment around them, and interpret its elements, according to their previous experiences and knowledge.

Interesting research by Gary and Wood (2011) discusses how not all mental models are created equal.[44] It seems it is quite hard to have accurate mental models of complex business environments, and it is much more rewarding and beneficial to have accurate models of key principles and causality relationships instead...It somehow goes back to the overarching principle of needing to live with incompleteness, not attempting to banish it.

Leaders who focus on trying to accurately map the business environment don't generate favorable results, as that effort might be diluted by complexity and the impossibility of predicting past the randomness. The latter option (focusing on guiding principles and causality) might help in making better decisions across a wider range of circumstances.

Both styles are approximations. However, one type of approximation leads to creative wins (leaders who keep a set of guiding principles in mind, and are open to different situations and flexible), while the other (leaders who try to approximate the environment according to their mental models) leads to illusions of control and knowledge, and to "too-confident" decisions based on deficient information.

Approximations that don't work

Many managers are bureaucrats who can end up wasting time and slowing down progress, without justifiable contributions. More managers than I care to remember serve as "email-relays." They get an email, and their utility (or lack thereof) lies in the delay each of them adds between the source of the email and its next recipient. The waste caused by "middle managers as email forwarders" is a persistent organizational nightmare for any business leader (Not everyone is an email forwarder. Many are meeting generators, spreadsheet populators, and presentation duplicators...Some of the rest (I think) are okay).

What are these low-performing corporate managers doing? Usually accumulating power and authority with a clear (and honest) end-goal: justify their existence a little bit longer. They habitually approximate work as the "appearance of work," or what is known as "staying busy."[45]

We approximate leadership as management, and management as leadership. We approximate work as the appearance of work...Appear busy, smile, and wave!

Eventually, and as this mix-up happens, and as more and more workers spend more time simulating work and producing evidence that they're working instead of really working, keeping themselves busy – at all costs, we have to ask ourselves what this does to the intrinsic and sacred value of real, creative, and productive work.

...also, to people's quest for meaning and growth.

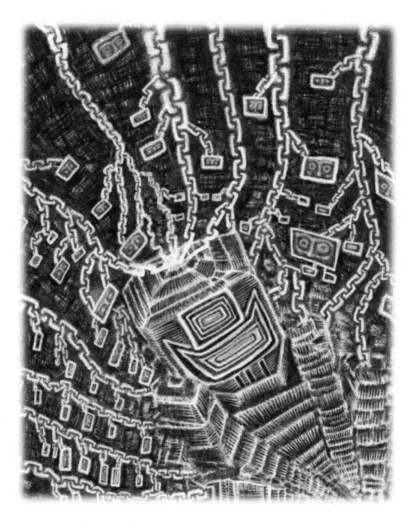

Summary: formula and examples

We make our life easier (almost) and use B instead of A…because B is more convenient, and/or A is more complex and difficult. Things are complex, our knowledge is incomplete, and we are lazy. We are what we were PLUS what we (approximately) learn/know/do every day, and we usually don't pay attention to all this.

In reality, $B = A + E$; where E is an error. E (fortunately/ unfortunately) is not completely random. Tragic outcomes can be expected when we have no idea of the approximation or estimate its scale and impact incorrectly. This happens because most of the time we are completely oblivious to the fact that we're approximating.

B is just fine (actually great) when we know about the approximation, because it makes solving problems and communicating easier. It allows us to venture into unexplored territories or attempt grand undertakings…

The act (?) of awareness itself transforms the fuzzy blade from a curse to a blessing!

This is the magical power of awareness, and its ability to transform is so elegant. It is sad that awareness and intentionality are absent most of the time.

The effects of being oblivious are sometimes tragic, and they span a wide set of situations and contexts, on both individual and group levels, and with regards to internal meanings and thoughts, as well as external behaviors, observations, and interactions.

In the next table I summarize a number of scenarios from the different fields, and dissect the exact approximations in them with a few comments.

The table shows the field, the thing we actually use because it is more convenient (B: the approximator), and the more elusive object which is intended and hidden (A: the approximated). As we do this, things are added or removed (approximated away), and these are contained in the error (E).

Note that A and B are interchangeable depending on the perspective.

Field	B (Apparent/Actual)	A (Hidden/Intended)	E (Error/Explanation)
Knowledge	Dr Slinky has two PhDs	Dr Slinky is knowledgeable and smart, and will know what to do	Dr Slinky might not know about a certain situation, and his/her knowledge might be a reflection of different and unneeded skills
Knowledge	Ahmad is a scientist	I trust him to decide for me	A scientist isn't an infallible super-human personality whose "expertness" (and integrity) in one field is transferable
Politics	The house of representatives has created 15 new laws in the past 2 years	The People's opinion is that these laws are fair and needed	Hidden agendas and implicit assumptions, along with divisions and "unknown unknowns" make it very dubious that E is small here. Most people might not know or care about these laws, let alone support them
Politics	A government welfare system is a good/bad thing	I am rich/poor so this affects me. My religious leader thinks that this is God's will	The issues we argue about are frequently just a veil for our identities and egos. Our group's values too as we ordain ourselves ultimate protectors of heritage
Management	Manager	Leader	An efficient person is no good replacement for someone who should break things and usher change
Management	Leader	Manager	A risk-taking and visionary person is no good replacement for an extremely conscientious and efficient organizer

Field	B (Apparent/ Actual)	A (Hidden/ Intended)	E (Error/Explanation)
Work	I sent 123 emails today (Activity)	I got the project finished. (Outcomes/ Actual Work)	Sometimes it is not important what precisely you did, but what you achieved
Language	The word: "Evolution"	Darwin, God, Survival, Atheism, Religion... (Set of related concepts and meanings)	Words are connected to networks of meanings and thoughts that are developed in the listener's mind. A discussion about evolution is not about evolution
Patriotism	Love for Country	Delusions of (personal) grandeur; Self-aggrandizement [46]	We are great, so I am great. Simple. Sometimes the opposite can be true too: My country is not good, but I am good, so I moved to this country which is good...like me. Please accept me as one of you, I promise to make fun of my (previous) country.
Decisions	"I think we should buy the new machine!" (A procurement recommendation)	Buying the machine is actually the optimal choice (Rational evaluation of alternatives)	We aren't, and maybe don't need to be, completely rational about decision-making...so let's stop pretending. I'm buying the machine because I want to grow my division, or impress my boss, or pay back the salesperson who is always nice to me.

Field	B (Apparent/ Actual)	A (Hidden/ Intended)	E (Error/Explanation)
Communicat- ion	"Statement 'A' is true." (My Ideas and Words)	I said that, and I am honest and I know what I'm talking about. You insult me if you refute it (Me / Myself)	We automatically assume certain positions. Even though ad-hominem is a common fallacy, it is almost always implicit: Your opinion is wrong → You are not smart!
Markets	The price of a pin is $0.1	$0.1 is the estimate of the Effort and Value of this pin	There are frequently inefficiencies in market-based pricing, making prices "unfair." Local conditions, information asymmetries, manipulation, and many other reasons can cause them. Still, price is usually considered to be a generally good approximator of effort and value (at scale)

* Hydra's Anger

(Or: Choose your weapon)

"It is probably because I don't know enough, but I can't shake off thinking that there should be a better way. Simply going on and just "building" the tower seems to be – in a sense – inefficient," Grumis said cautiously, with the intent of probing without affecting the work.

"Yes. That's what you'd think as I told you before. I thought that too many times. Too many times. This is how we're doing things now, and it is working. Maybe not as it should, but I can't really complain. It takes so much time to change a simple thing in everyone's work mode, that makes the whole thing not worth the effort."

"You mean if you tried to impose some coordination on them?"

"Well...Paloneol was a colleague of ours, and he once said, 'I know what the tower will look like...I can see it in my mind's eye, and it is so beautiful and majestic.' He would always talk about its final shape, and get engrossed in his notebook of scribbles. When we asked him where to put that stone, he wouldn't know, but he would then get angry after a while claiming that things are not being done correctly. He would sometimes shout comments that no one else could understand or do: 'get me a dodecahedron rock' he would shout...'We don't have it,' 'nonsense!'...and on the discussion would go."

"And what happened...?"

"One of his colleagues threw him from the tower. Haven't seen him since. Lanoleop was his name. Lanoleop promised me that he knew what needs to be done. He would tell every worker which rock to carry and where to place it. It was somehow annoying for them. Most annoying though was that the tower turned out – after a few months – to be leaning, and poor Lanoleop fell too. Haven't seen him since."

"So neither style worked?"

"You see, telling them what to do, and how, just didn't work. We deeply wished we'd be able to solve the problems by figuring them *out in advance, but no matter where we focused, we were surprised with new classes of problems. Also, when you've done something for a while, you don't really want someone telling you how to do it, do you?"*

"I think so..."

"Tell them about the rocks, the wood becomes a problem. Tell them about working in small units, anger becomes a problem. Tell them about working on a specific area, the other becomes a problem. Tell them to look closely, they fall...One problem solved causes three others to rise. Like that monster...The Hydra."

"And does everyone want to do things like this...?"

"That's not important. We've learned not to dwell too much on what everyone wants. If anyone says something, or tries to change something, we can easily get into a fight that will get almost everyone here deleted. Groups will form easily, over completely trivial differences, and they will fight with ferocity you've never seen."

"Fighting over changing the work method...?"

"Not really. Fighting over anything, including the work method. One time the workers fought over whether bigger left ears were more useful than bigger right ears. Many died. Those who remained fought over whether equally-sized ears were more useful than differently-sized ears. It really doesn't matter once you start fighting."

"So the only solution..."

"Is to just build the tower!"

Notes

1. The AQAL framework was developed by Ken Wilber and is discussed in many books either by him or by others that followed the framework. Example: his 2001 book: *A Theory of Everything: An Integral Vision for Business, Politics, Science and Spirituality.*

2. Wilber (2000) discusses two types of reductionism, the extreme form of which reduces everything into the Objective-Individual (atomistic) realm, and the subtle form reduces the whole collective reality to the left-hand side (exterior), so social systems are assumed to be comprehensive at the expense of understanding how inner-collective meanings emerge and are experienced.

3. Even though Nietzche might disagree with this claim: "Man (?) does not strive for happiness; only the Englishman does that."

4. Refer to the article by Shah & Oppenheimer (2008).

5. More on these Biases and Heuristics in Simon (1990).

6. See *The Cognitive Bias Codex* by DesignHacks.co as a nice example

7. Refer to the research by Heck, Simmons, and Chabris (2018) for more on the egocentric bias.

8. Refer to the article by Pronin, Lin, and Ross, (2002).

9. For more on the science of sports fans see the study mentioned here: https://www.newswise.com/articles/testosterone-levels-rise-in-fans-of-winning-teams , by Bernhardt from the University of Utah.

10. Daniel Wann, a social psychologist at Murray State University, has done considerable research on these phenomena. There is also a book he's co-authored with Jeffrey James (2018) on the topic.

11. Barth (2016) "Stadiums and Other Sacred Cows: Why questioning the value of sports is seen as blasphemy."

12. W.B. Yeats in his poem: "The Second Coming" (1920).

13. Refer to the article by Handel and Schawartzstein (2018).

14. For a detailed discussion on brands refer to Wood (2000).

15. The same article by Handel and Schawartzstein (2018),

16. Just as some psychology research tries to define "dimensions of personality" for humans, research has developed similar "personality dimensions" models for brands, like the work of Aaker (1997), which says that a brand's personality can be defined by its sincerity, excitement, competence, sophistication, and ruggedness. This applies to American brands/consumers, and changes in culture will lead to changes in the model.

17. Refer to the article by Fitzsimons, Chartrand, and Fitzsimons (2008) for a detailed discussion on how this experiment of "priming with apple," then measuring creativity was conducted.

18. This is discussed by Saussure (1959). Note that neither the signifier nor the signified are actual things: one is an external/collective artifact, the other is an internal/individual understanding.

19. There are many hilarious celebrity (especially sports' celebrities) endorsements. Look them up (hint: Shaquille O'Neal and Cristiano Ronaldo, among others, have done more than 30 endorsements each).

20. Dentists seem to be finding it hard to agree on one toothpaste/mouthwash (too many)...Doctors are more decisive: "More doctors smoke Camels than any other cigarette...The doctors' choice is America's choice," goes the popular meme-ad.

21. The formal sciences are usually concerned with symbols and rules that model their relationships. Examples of applied sciences include engineering and management sciences.

22. Even though I didn't go into the –isms in detail in the text, I thought that including some in a footnote might be fun for some* people. Here is a list of -isms relevant to science: Rationalism, Empiricism, anti-rationalism, Inductivism, Idealism, Skepticism, Functionalism, etc…look them up and have fun!

23. The logic and terminology in the discussion of evolution of science mainly comes from the work of Thomas Kuhn, "The Structure of Scientific Revolutions."

24. Bernhard Riemann (1826–1866) was a German mathematician whose work in differential geometry helped lay the foundation for general relativity (he had numerous other contributions too).

25. Refer to Keith Stanovich (2007).

26. The audacity that some researchers have, to compare financial results of companies across geographies, industries, or time, and then relate that to strategic or organizational variables, is one of the world's wonders and mysteries.

27. Science as Ego and Politics: The article "The Politicization of Science" by Jeanne Goldberg (2017) provides a good history and overview of different ideas mentioned in this section.

28. Max Planck (1949) Scientific Autobiography and Other Papers. He also said: "Science advances one funeral at a time."

29. It becomes a vicious loop of publishing for publishing, because this is what you do now. Many "old and respected" professors end up having their names on 100+ publications, by means of reproduction, mediocrity, and coercion of young researchers to include their names in the author list in return for dubious theoretical contributions or the prestige of association.

30. Charles Murray's book *Coming Apart* (2013) shows a direct correlation between years of education completed and

IQ, meaning that PhD-holders (20+ years of education) have the highest average IQ as a group compared to other segments as grouped by years of education completed. The reader can also refer to the work by Dutton & Lynn (2014).

31. This quote is from the book *Bushido – the soul of Japan*, by Inazo Nitobe. It was written more than a hundred years ago, so we won't be overly critical.

32. Refer to the work of Capra (1997).

33. For the interested champions of knowledge (nerds), check the discussions of the "etic" vs "emic" approaches in anthropology.

34. Languages can color reality, sometimes literally, as some research has shown that certain people (the "Himba of Namibia") who didn't have the word "blue" in their language, couldn't easily tell the difference between blue and green (Roberson et al, 2006). A popular internet quiz does the same for English speakers and different shades of green.

35. The weak form of the Sapir-Whorf proposition, which says that language influences (and not determines) thought patterns has produced positive empirical results (Ahearn, 2011). (Note: treat "empirical results" in the social sciences with caution.)

36. Nerd Alert: more accurately, they do this by eliminating syntactic ambiguity, and minimizing semantic ambiguity, if you're into the specifics. In other words, removing the excess meanings and extra colors of words and phrases.

37. Gismu is Lojban grammar. https://www.lojban.org/files/why-lojban/whylojb.txt

38. A formal description of these wicked problems can be found in Rittel and Webber (1973).

39. Look for this in (somehow) democratic countries like France or the US, because – statistically – people seem to be happier when ruled by a (visible) ruthless dictator, or

when there are no freedoms of the press. Consult work by "free" journalists, country-ranking agencies, and your local reputed think-tanks and research centers to verify.

40. Refer to the Book: The New Leaders (2002) By Boyatzis et al. Around 4.5 million free copies of the book were distributed for free by 2020 by the Ayn Rand Institute (https://issuu. com/aynrandinstitute/docs/237692_aynrand_r2_proof). For more details refer to Doherty (2007).

41. Full disclosure: Watched at 1.25x, and it was still a bit too slow. Also: movies are almost never a good replacement for a book.

42. Ayn Rand is reported to have said that "the only philosopher to have influenced her thinking was Aristotle," and I – automatically – thought how his assertion that women have fewer teeth was quite John-Galtian.

43. As always, I remain very skeptical of this kind of strategic/managerial research, so I include this example here anecdotally and not to be taken as a "scientifically-supported" hypothesis...

44. There is actually a book called *Bullshit Jobs: A Theory* (2018). I prefer a more nuanced terminology.

45. Scott wrote of patriotism, that "as it is the fairest, so it is often the most suspicious, mask of other feelings."

The Creative Escape: Dart the Dark Side

Your assumptions are your windows on the world. Scrub them off
every once in a while, or the light won't come in.

Alan Alda

In this short section, I attempt to shift the discussion into a more prescriptive one.

Far from claiming the role of the wise man (or healer), I try to use common sense and some insights from the journey and different disciplines to come up with a set of recommendations. You don't need to trust them, but they add to the discussion, and might help alleviate a few of the problems caused by approximation.

Remember its great positive and constructive potential!

Look at the bright side (and then at the dark side again)!

No matter how thin you slice it, there will always be two sides.

Baruch Spinoza

Also, look in the mirror.

The insights from quantum mechanics and evolution (as super ideas) can be applied to many different fields. Their applications extend to the cognitive and social fields with stunning elegance.

Just like different co-existing possibilities and versions of reality are separated by an observer's (mean) observation intervention, and collapse into one reality, so too the different faces of approximation are separated by a moment of awareness.

I call it a moment of awareness because it seems to me that it is less about the knowledge than it is about the will.

A moment of awareness is the difference between the bright and the dark sides of approximation, and can draw the line between open exploration and growth on the one side, and closed defensiveness and calcification on the other.

Living with uncertainty is inescapable because as we saw, our understanding will always be a work in progress. To live up to a destiny of being truly creative, one must embrace the ambiguity and accept risk (to the ego/self and other things).

Understanding is incomplete, and it is muddied by emotions, preferences, and perceptions. Reality is complex, non-linear, and out of control. Relying on fuzzy and approximate ideas can transform paralysis to progress, but the caveat is not to get too attached to a limited view of the world, group, or self.

There is no growth without accepting the current state of incompleteness, and being ready to abandon it for a new position.

That being said, extending the time between perception and reaction might not be a simple task. Sometimes reality can be thin, and its two faces too close.

"Be aware" is not proper advice, and it can't really work without necessary adjustments in character and circumstances.

A moment of awareness is (to use a cliché) the tip of the iceberg. It hinges on a lifetime of responsibility and character evolution. Responsibility starts with taking ownership and shifting the gaze (and blame, more than credit) to the inside.

The individual is powerful precisely because of this power over destiny...and personal destiny is also universal destiny.

The problems that arise from uncaught approximations are not likely to disappear any time soon but it might be entertaining and informative to wrap things up with a look at a few recipes that can make a moment of awareness more likely.

Many of the ideas here are also solutions to other problems (because – as I said before – approximation might be an overlooked super idea), so recounting them in this context might provide further motive to embrace them (you've probably seen them before, so the following discussion will make them more memorable, and increase the likelihood they "pop-up" into your internal monologue).

Incremental and sustainable growth: soft things are actually tough

Solving the problem of "bad" approximation is no easy task. Being a "wicked" problem itself, it involves complexity, is affected by emotions, and relates to other problems. Some of its impacts can be remedied by taking a long-term perspective.

Paying attention to instances of defensiveness (as they happen, and later) can be a great habit.

When did you cling to what exists – as-is – so stubbornly that it prevented growth or caused conflict (as bad as that "other person" is)?

With time, things that are unacceptable or that can't be clarified by theories and words will become evident to the person who looks. Why else would experience be considered the best teacher (even if it is a highly inefficient one)?

Many of our struggles and stubborn confrontations are a desperate effort to protect some self or image of the self, which frequently isn't worth – or doesn't need – protection.

> *"Know Thyself"*
> *"The unexamined life is not worth living."*
>
> Socrates

Observing is always THE first step.

As some of these motivations are examined, one becomes more tolerant and gets a "thicker skin." This means that defensiveness starts fading away, leaving room for growth. If you are not really in danger (real or perceived), there is no reason to be furious!

We tend to live in compartments, sometimes too small. It is easier to divide the world, so we tend to define ourselves in terms of categories that are sometimes too limited. We approximate ourselves as this or that, and so we are naturally prone to confusing concepts and defending different things because the boundary between them and our identity is blurred.

You are not a student, sales manager, engineer, or penguin-straightener (A penguin-straightener's role in a zoo is to help penguins who have fallen and can't stand up. This requires further research).

These are tasks you do. They could become who you are, but that is a choice and a potentiality, not a destiny. You are not who you think you are – this is temporary and evolving.

You are not just a part of your group – that too can change.

You are more accurately (accuracy is relative, so this claim is still an approximation) described as a symbol-manipulating consciousness (a recursive one, no less[1])...your symbol set is (supposed to be) always growing and evolving.

When you define yourself as someone in a box, you might take many things for granted that others don't. This is a great cause for unseen approximations. Pay attention to when you say "us," and think about the discussion on our tendency to "bask in the glory" and attribute some vague greatness to ourselves via these means.

Try to grow beyond the social box that has been placed (by you or others) around you!

This also means that you must try to use and engage with new and different cultural tools and artifacts. Meet new people, read different things, listen to new music, go outside the work-prison...

One great way to do that is the "humanities" knowledge field. The arts are very important to your evolution as a person. Read good literature, watch good films, listen to good music, (or bad ones, if you prefer, what's important is that you practice judgment) and this will help you see perspectives in the world that you don't have time (or energy) to explore yourself.

A reader lives a thousand lives before he dies, said Jojen.
The man who never reads lives only one.

George R.R. Martin [2]

Life is too short for one box.

Don't dismiss philosophy altogether as something useless and impractical.[3] It will do wonders to your thinking, which you can take to other fields of life. Find what interests you, and start there.

Emotional tendencies to defend certain positions can be clarified when – in an argument – one plays the role of the "other side." By all means, be your own devil's advocate (remember Dunning-Kruger).

In this sense, it is definitely valuable to learn about – and develop – the different skills that fall under the "emotional intelligence" umbrella. Even though some might object to the rigor (provable scientific validity) of the claims of the proponents of the field, I've found the development of some of these skills very valuable personally and professionally. Daniel Goleman's landmark book (*Emotional Intelligence*, 2005) is definitely a great place to start.

If you're worried these skills are too wasteful or too passive, don't be. All these "soft" activities will not necessarily make you soft. They are – in fact – a great repertoire for success in our modern shallowly analytical world. The combination of the above with a clear and conscientious disposition, a knowledge of your practical field, and an intention for impactful contribution, is what the world now desperately needs (even/especially professionally).

Learn from designers

Design thinking is a set of processes that include different cognitive and managerial tools to develop new concepts and build/generate new products and experiences (the version applied to management). Ultimately, design thinking is about understanding, empathy, and creative problem solving. It often centers around dealing with ambiguity and uncertainty.

I think that some lessons from design thinking can apply to the general formulation of problems that are caused by unseen approximations.

Design thinking tries to reframe design problems to put the user first. It looks for insights through observation and the mapping of the experiences of others. It requires the designer to reimagine the product and reconfigure it if needed, breaking away from the inertia and momentum that have been stuck there by time and tradition.

Understanding users and their experiences is crucial, and so is seeing the world from their perspective. This is a kind of empathy and understanding that compels one to go beyond his/her own set of mental representations, and can thus lead to insights about the world and about other people's experiences. In a sense, it advocates breaking free from the limitations of one's own perspective, and from the existing norms.

It is a search and refinement of approximations.

The immersion of the designer, or any person attempting problem solving, into another person's experiences enriches their perspective and helps in the discovery of previously hidden approximations. This is usually done through a planned and deliberate set of actions that generate empathy and understanding, and include open discussions, observation, and mimicking experiences of others in what is known as a "deep dive."

This empathy is frequently coupled with a simplification, but from the new perspective.

Instead of the previous approximation, a new – and more creative – approximation is generated. Situations and tasks are sometimes simplified and experienced in order to understand core concerns.

Designers need to live with uncertainty and must acknowledge the incompleteness of knowledge. One of the most memorable thoughts for me from early design management classes was how uncomfortable the fuzziness of the experience in the first few sessions felt. I – being used to rigorous problem solving at the time – absolutely hated the loosely defined problems and the absence of clearer requirements and frameworks to be used in reaching solutions (I also hated the many post-it notes hanging everywhere).

The trick is to constantly – and practically – test the limits of your knowledge and scope. Admit ignorance, and the possibility of failing. The activity constantly exists on the border between the known and the unpredictable, and this awareness of the incomplete perspective leads to openness and acceptance, and to an appreciation of ambiguity in the context of solving problems.

Design thinking is always coupled with continuous education, seeking a breadth of knowledge across disciplines, while having deep specialized knowledge within a certain field. People whose knowledge and expertise are T-shaped (so – just like the letter "T" – broad, but with extensive specialization in one area) have long been praised in design-oriented teams because they can help build a collaborative and evolving culture, and solve a diverse set of problems as team members[4].

Just in case you're wondering about the "T," there are also I-, Pi-, M-, E-, and X-expertise people. So how are you shaped?[5]

Practice every day (hour?)

The set of recipes presented here is not large because it is not the point of the book. Much more interesting is the exploration of this approximation force that stealthily influences different parts of our social and intellectual lives. I'd like to conclude the chapter with a set of observations derived from personal experiences and reflection.

We need to accept the fact that understanding has a magical and fleeting component. Understanding is like holding something in mind, when you remove your focus, it will be lost – as if vanishing in the ether. It is important to note, however, that grasping it becomes significantly easier with every new attempt. The same goes for mental habits. This relates to the default-mode network we discussed before, so again, one must be very careful about what ideas get deposited there. Things start consciously, but acquire a life of their own.

I begin with an idea and then it becomes something else.
Pablo Picasso

Be explicit about unknowns whenever you can (when solving problems, discussing, or even just thinking). This is something that consistently gets neglected. Remember the large ocean of "unknown unknowns": acknowledging it can be a source of limitless potential. Try to prod others into seeing their unknowns, but only after you've done so yourself – it is easier that way.

Be skeptical of analogies in discussions. They are the secret tunnel that can bring too many unwanted – and disguised – attendees to the table. They can also bring many insights and allow for a better understanding of the different perspectives.

If you do get those, try to move to the same layer/plane your counterparts are coming from (remember projection).

One must always remember the hidden cages that are created by groups. Groups are great sources for creativity and survival, but like approximation, they have a dark side. Their pressures can sometimes mask many biases or enforce specific ways of thinking. This can be countered by conscious and critical reflection, and the search for the "comfort in numbers" tendency.

Taking risks to grow, owning the processes of growth, and embracing creativity are the human destiny. Groups tend to do that collectively and individuals often pay the price (statistically speaking). Try to be among the lucky (sub)group.

All the above opportunities for reflection usually present themselves (at least) on a daily basis – I encourage you to examine them with the tools we've discussed, and to reflect on incompleteness, awareness, attention, knowledge, and the borders of your "self."

Don't take culture "as-is"

One of the important insights on culture is that it can work as a scaffold for supporting creativity and progress, because it provides certainty and confidence when they are most needed. It can also synergize different mental and emotional energies. However, we've seen that culture includes a set of values that relate the different (visible) elements, and deeper still, a set of assumptions that most people remain unaware of[6]. This "unawareness" is precisely a cause for alarm. Culture is the source of many approximations that we systematically repeat and generally take for granted.

Is a certain cultural affiliation (remember, you have many) limiting you, and causing "tunnel vision"? Is it driving you toward conflict? In fact, is it suitable for thinking and acting within a certain context you're in? Are its "resources" enough or suitable for what you're trying to achieve?

Reflecting on our "systems of shared meanings" (and actively selecting to change them) is an essential responsibility.

What do I value? Why? Why is that thing beautiful?... Meaningful?...Just?

Trying to trace how the meaningful symbols of our life relate to our deep and unexamined assumptions is an overlooked treasure chest of regeneration and growth!

* Syz of Zif

(Or: The unseen reflection)

"I want now – if you don't mind – to ask you the most important question I've been thinking about since I came here," Grumis said, recording and monitoring Zif's every feature. "What would that be?"

"Well…why are you building this tower?"

"Ha! How's that important? Isn't it obvious? I never got to ask you where you come from."

"It is not obvious to me."

"Syz. Syz told me to build it."

"So what he wants is important?"

"Yes. I'm doing what Syz wants. Everyone does what Syz wants."

"Oh, and when did you see him? I mean did he tell you to build this?"

"Not really, no. But everyone knows what he wants."

"But why? Who is he?"

"We don't really have time for that…Why the tower? Who is Syz? As you see we have so many unfinished tasks and difficult obligations. Who has time to go on strange quests of philosophical nature? Maybe later, when we've done the biggest part of the work, we'll have the luxury of being able to look into these inconsequential matters."

"How will you know it is the time for learning about Syz?"

"What I know seems enough for now. Things are, and it is simpler this way."

"I see," said Grumis.

But he didn't see. Zif is stuck in a non-ending loop.

Where would Zif's quest for finding Syz lead?

What drives his solid motivation?

Shouldn't that have been his task, before moving rocks?

Why didn't he even try to grasp the specific reality of Syz and his wishes?

Importantly, what if Syz was a voice in his own self, telling him to move the rocks, because he was already moving the rocks?

As significant sections of the tower came tumbling down, Grumis started the sequence for exiting this world. He recorded the different variables. He noted how almost all the encounters with incompleteness had turned out to be negative here.

He tried to survey the reasons and record them.

In other worlds things could have been easier to survey, there could have been less variability, easier communication, and even more stable and friendly solids. People could have been more detached and knowers could have been more cooperative, and – well – knowing.

What if a few variables were adjusted? He understood that flipping a few variables didn't necessarily mean a better fate for Zif, but the sense of the experience could have been drastically different.

Zif, Syz, the workers, and all the others were forever marked in him now. His neural networks were updated to include their experiences and thoughts.

Grumis logged out, as Zif kept doing what he has been doing for as long as he remembers.

Building the tower.

Notes

1. Refer to the beautiful book by Hofstadter (2008), *I Am a Strange Loop*.
2. This quote is from his *A Dance with Dragons*.
3. Sadly, being "philosophical" is a derogatory comment sometimes synonymous in spoken Arabic with being impractical or wasteful. Probably in other languages too.
4. IDEO's CEO Tim Brown: https://chiefexecutive.net/ideo-ceo-tim-brown-t-shaped-stars-the-backbone-of-ideoaes-collaborative-culture
5. Refer to this: https://www.axelos.com/news/blogs/august-2020/what-employee-type-or-shape-are-you
6. Refer to Schein's (2006) work mentioned earlier.

Epilogue: Eternity in a Moment

Even though the book discusses approximation and fuzziness, the central theme that underlies the whole discussion is the inevitable and perpetual incompleteness of the human experience.

By discussing the elements that make it inevitable in our cognitive and behavioral lives, and examining the everyday life problems that arise from our abuse of the very tools that help us deal with incompleteness, I was actually trying to paint an inverse picture.

It is a more optimistic picture that centers around the creative potential built into incompleteness and our interactions with it. This incompleteness is what opens the door for change, and we frequently forget that as we resort to our safe bunkers.

Think about it: the creative destiny of humanity is nurtured by this all-encompassing, somehow subjective, incompleteness. The creative destiny elegantly rests on one moment. It takes just one moment of awareness to define the difference between the two faces of approximation.

A moment of intent and humble reflection stands between the creative mists of progress, and the prisons of illusion, waste, and conflict.

Carry the burden of responsibility as you (continually) cross from the dark side.

Bibliography and Further Reading

Aaker, J. L. (1997). Dimensions of Brand Personality. Journal of Marketing Research, 34(3), 347–356. https://doi.org/10.1177/002224379703400304

Abbott, E. (1884) *Flatland.*

Ahearn, L. M. (2011). *Living language: An introduction to linguistic anthropology.* Wiley-Blackwell.

Andrews-Hanna, Jessica R. (2012-06-01). "The brain's default network and its adaptive role in internal mentation." *The Neuroscientist.* 18 (3): 251–270. doi:10.1177/1073858411403316

Barth B.J. (2016) "Stadiums and Other Sacred Cows: Why questioning the value of sports is seen as blasphemy." Nautilus. August 2016. http://nautil.us/issue/39/sport/stadiums-and-other-sacred-cows

Boyatzis, R.E., Mckee, A., and Goleman, D.J. (2002). *The New Leaders: Transforming the Art of Leadership into the Science of Results.* Sphere.

Capra, F. (1997) *The Web of Life: A New Scientific Understanding of Living Systems.* First Anchor Books.

Dawkins, R. (2008) *The God Delusion.* Bantam Press

Doherty, Brian (2007). *Radicals for Capitalism: A Freewheeling History of the Modern American Libertarian Movement.* New York: Public Affairs Press

Dunning (2011) The Dunning-Kruger effect: On Being Ignorant of One's Own Ignorance. Advances in Experimental Social Psychology. 44

Dutton & Lynn (2014) Intelligence and Religious and Political Differences Among Members of the US Academic Elite (Interdisciplinary Journal of Research on Religion): http://www.religjournal.com/pdf/ijrr10001.pdf

Garnot, Y., Balcetis, E., Schneider, KE, Tyler T.R. (2014) Justice is not blind: Visual attention exaggerates effects of group identification on legal punishment. Journal of Experimental Psychology: General

Gary MS and Wood RE (2011). Mental Models, Decision Rules, and Performance Heterogeneity. Strategic Management Journal, 32

Godin, S (2020). *The Practice: Shipping Creative Work*. Portfolio.

Goldberg, Jeanne (2017) The Politicization of Science by Jeanne Goldberg. Skeptical Inquirer Volume 41.5, September/October 2017

Goleman, D. (2005) *Emotional Intelligence: Why It Can Matter More Than IQ*. Bantam Books.

Gráinne M. Fitzsimons, Tanya L. Chartrand, Gavan J. Fitzsimons, Automatic Effects of Brand Exposure on Motivated Behavior: How Apple Makes You "Think Different," Journal of Consumer Research, Volume 35, Issue 1, June 2008, Pages 21–35, https://doi.org/10.1086/527269

Handel B, Schawartzstein J (2018) Frictions or Mental Gaps: What's Behind the Information We (Don't) Use and When Do We Care? Journal of Economic Perspectives 32:1 pp: 155-178

Heck PR, Simons DJ, Chabris CF (2018) 65% of Americans believe they are above average in intelligence: Results of two nationally representative surveys. PLoS ONE 13(7): e0200103. https://doi.org/10.1371/journal.pone.0200103

Hijazi, A., Sinha, S. (2020) On Ethereal Grounds: Cultural Resources as Foundations Supporting Innovation Success. Journal of International Consumer Marketing. DOI:10.1080/08961530.2020.1798837

Hofstadter, D. (2008). *I Am a Strange Loop*. Basic Books

Kelp, C. (2015). "Understanding Phenomena". Synthese. 192:12 pp:3799-3816

Killingsworth, M. A., Gilbert, D. T. A. (2010). Wandering Mind is an Unhappy Mind. Science.

Kruger and Dunning (1999) Unskilled and Unaware of It: How difficulties in recognizing one's own incompetence lead to inflated self-assessments. Journal of Personality and Social Psychology. 82

Kuhn, T (1962) *The Structure of Scientific Revolutions*. The University of Chicago Press.

Levitt, Steven and Stephen J. Dubner (2005). *Freakonomics: A Rogue Economist Explores the Hidden Side of Everything*. William Morrow/HarperCollins

Lewandowska-Tomaszczyk, Barbara (2012). Approximative Spaces and the Tolerance Threshold in Communication. International Journal of Cognitive Linguistics 2(2). 2–19.

Marcus E. Raichle, Ann Mary MacLeod, Abraham Z. Snyder, William J. Powers, Debra A. Gusnard, Gordon L. Shulman. (2001). A default mode of brain function. Proceedings of the National Academy of Sciences Jan 2001, 98 (2) 676-682; DOI: 10.1073/pnas.98.2.676

Murray, Charles (2013). Coming Apart: The State of White America, 1960-2010. Crown Forum; Illustrated edition (January 29, 2013)

Pronin, E.; Lin, D. Y.; Ross, L. (2002). The Bias Blind Spot: Perceptions of Bias in Self Versus Others. Personality and Social Psychology Bulletin. 28 (3): 369–381

Rittel, Horst W.J.; Webber, Melvin M. (1973). Dilemmas in a General Theory of Planning (PDF). Policy Sciences. 4 (2): 155–169. doi:10.1007/bf01405730

Roberson, D., Davidoff, J., Davies, I. R. L., & Shapiro, L. R. (2006). Colour categories and category acquisition in Himba and English. Volume II. Psychological Aspects, 159–172. https://doi.org/10.1075/Z.PICS2.14ROB

Sarkar, Christian and Kotler Philip (2020) *Brand Activism: From Purpose to Action*. Idea Bite Press

de Saussure, Ferdinand. *Course in General Linguistics*. (1959). The Philosophical Library, New York City.

Schein, E. H. (2006). *Organizational Culture and Leadership*. Jossey-Bass.

Shah & Oppenheimer (2008) Heuristics Made Easy: An Effort-Reduction Framework : Psychological Bulletin: 2008, Vol. 134, No. 2, 207-222

Simon, Herbert A. (1955-02-01). A Behavioral Model of Rational Choice. The Quarterly Journal of Economics. 69 (1): 99–118. doi:10.2307/1884852.

Simon, H. A. (1990). Invariants of human behavior. Annual Review of Psychology, 41, 1–19.

Stanovich, Keith E. (2007). *How to Think Straight About Psychology.* Boston: Pearson Education

Taleb, N. (2012) *Antifragile: Things that gain from disorder.* Random House.

Tajfel H, Turner J. (2004). *Political Psychology. Chapter: The Social Identity Theory of Intergroup Behavior.* Imprint Psychology Press

Thompson D (2017) Hit Makers: The Science of Popularity in an Age of Distraction. Penguin Press

Thorndike, E.L. (1920). A constant error in psychological ratings. Journal of Applied Psychology, 4(1), 25–29. https://doi.org/10.1037/h007166 3

Vanderbilt T. (2014). How your brain decides without you. Nautilus. 19 (Illusions).

Wann, D. James, J. (2018). *Sport Fans: The Psychology and Social Impact of Fandom.* 2nd Edition. Routledge.

Wilber, K. (2000). *Sex, Ecology, Spirituality: The Spirit of Evolution.* Shambhala.

Wilber, K. (2001). *A Theory of Everything: An Integral Vision for Business, Politics, Science and Spirituality.* Shambhala

Wood, L. (2000) Brands and brand equity: definition and management, Management Decision, Vol. 38 No. 9, pp. 662-669. https://doi.org/10.1108/00251740010379100

Zadeh, L.A. (1965). "Fuzzy sets." Information and Control. 8 (3): 338–353

Zajonc, Robert B. (1968). "Attitudinal Effects Of Mere Exposure."
Journal of Personality and Social Psychology. 9 (2, Pt.2): 1–27.
doi:10.1037/h0025848

IFF
BOOKS

ACADEMIC AND SPECIALIST

Iff Books publishes non-fiction. It aims to work with authors and titles that augment our understanding of the human condition, society and civilisation, and the world or universe in which we live. If you have enjoyed this book, why not tell other readers by posting a review on your preferred book site. Recent bestsellers from Iff Books are:

Why Materialism Is Baloney
How true skeptics know there is no death and fathom answers to life, the universe, and everything
Bernardo Kastrup
A hard-nosed, logical, and skeptic non-materialist metaphysics, according to which the body is in mind, not mind in the body.
Paperback: 978-1-78279-362-5 ebook: 978-1-78279-361-8

The Fall
Steve Taylor
The Fall discusses human achievement versus the issues of war, patriarchy and social inequality.
Paperback: 978-1-78535-804-3 ebook: 978-1-78535-805-0

Brief Peeks Beyond
Critical essays on metaphysics, neuroscience, free will, skepticism and culture
Bernardo Kastrup
An incisive, original, compelling alternative to current mainstream cultural views and assumptions.
Paperback: 978-1-78535-018-4 ebook: 978-1-78535-019-1

Framespotting
Changing how you look at things changes how you see them
Laurence & Alison Matthews
A punchy, upbeat guide to framespotting. Spot deceptions
and hidden assumptions; swap growth for growing up. See
and be free.
Paperback: 978-1-78279-689-3 ebook: 978-1-78279-822-4

Is There an Afterlife?
David Fontana
Is there an Afterlife? If so what is it like? How do Western
ideas of the afterlife compare with Eastern? David Fontana
presents the historical and contemporary evidence for
survival of physical death.
Paperback: 978-1-90381-690-5

Nothing Matters
a book about nothing
Ronald Green
Thinking about Nothing opens the world to everything by
illuminating new angles to old problems and stimulating new
ways of thinking.
Paperback: 978-1-84694-707-0 ebook: 978-1-78099-016-3

Panpsychism
The Philosophy of the Sensuous Cosmos
Peter Ells
Are free will and mind chimeras? This book, anti-materialistic
but respecting science, answers: No! Mind is foundational to
all existence.
Paperback: 978-1-84694-505-2 ebook: 978-1-78099-018-7